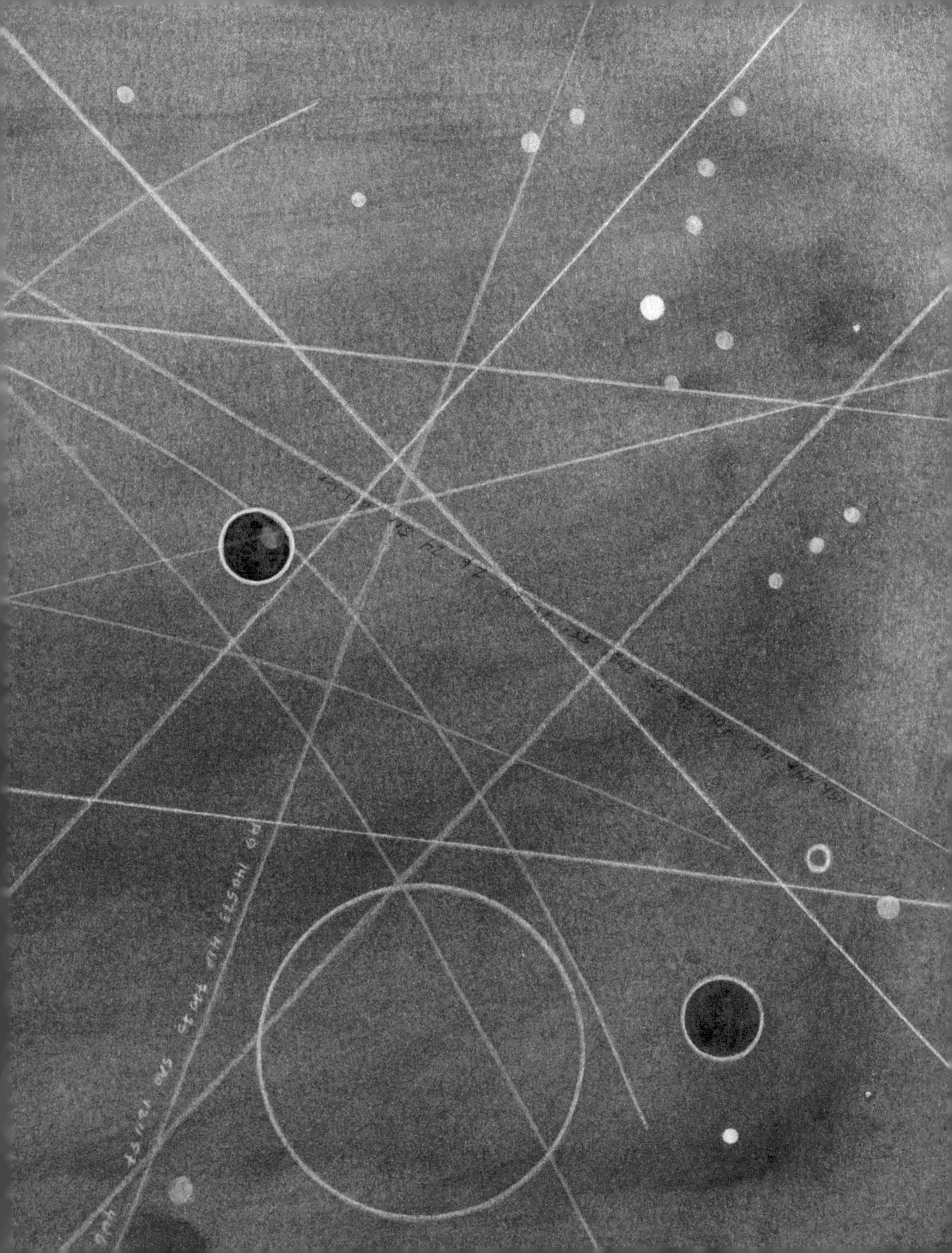

SPACE

A COLLECTION OF ESSAYS AND IMAGES

SHANA MABARI & ANDI CAMPOGNONE

Griffith Moon

Published by the Lancaster Museum of Art and History (MOAH) on the occasion of the exhibition The Light of Space, February 8 – April 19, 2020

Printed in the United States of America
First Printing, 2020

ISBN: 978-1-7326992-9-8
Library of Congress: 2020904003

Published by Griffith Moon
Santa Monica, California
www.GriffithMoon.com

"Children at play, artists at work, and scientists measuring quantum effects share this in common: they are all creating reality."

- LEONARD SHLAIN

CONTENTS

FOREWORD

There is geometry in the humming of the strings, there is music in the spacing of the spheres.

– Pythagoras

It is no coincidence that the Light and Space movement of the 1960s and '70s originated in Southern California, in proximity to the burgeoning aerospace industry, which remains an enduring presence. The movement emerged as a direct interpretive response both to the broader cultural obsession with space exploration and to the specific ways in which it was altering our species's sense of time, scale, geometry, and perception. The fundamental restructuring of humanity's sense of itself was as irresistible to artists, writers, and filmmakers as it was to scientists, engineers, and mathematicians.

Beyond the zeitgeist, this fascination also had a remarkable material aspect, in which not only ideas and perceptual frameworks were borrowed from scientists for art-world purposes, but also space-age materials such as resin, fiberglass, telescopic lenses, and unconventional light sources.

This book amply reflects that spirit of collaboration and appropriation, both intellectual and material. Shana Mabari's inspired idea to curate a selection of essays on the concept of space as part of her contribution to *The Light of Space* exhibition at MOAH originated in her own career-long exploration of the intersections of art and science, most specifically, in 2018, as she was preparing to become the first artist to fly on a mission of NASA's Stratospheric Observatory for Infrared Astronomy. Her invitation to others from diverse professional backgrounds to contribute texts has yielded a collection that gives dazzling prismatic effect to the otherwise ordinary term *space* while at the same time reminding us of the interdisciplinary foundations of the Light and Space movement. The essays alternate with the work of contemporary artists who represent the movement's materials and values, and who are carrying forward its legacy into new generations, resonances, concerns, and contexts.

Across fine art, neuroscience, cognition, evolution, emptiness, place, metaphor, energy, distance, pictorialism, ontology, poetry, and myth—across sculpture, painting, photography, musicality, rocket science, architecture, astronomy, abstraction, color theory, atmospheric light, and fractal mathematics—the scope of this inquiry sails from the quantum to the conceptual, and from the intimate to the intergalactic, all in the service of a deeper understanding of the universe and our place in it.

Thank you to everyone who has made this voyage of exploration possible. Your open minds and ambitious visions will continue to inspire here on Earth as long as there are stars in the sky.

- Andi Campognone
Curator, Lancaster Museum of Art and History

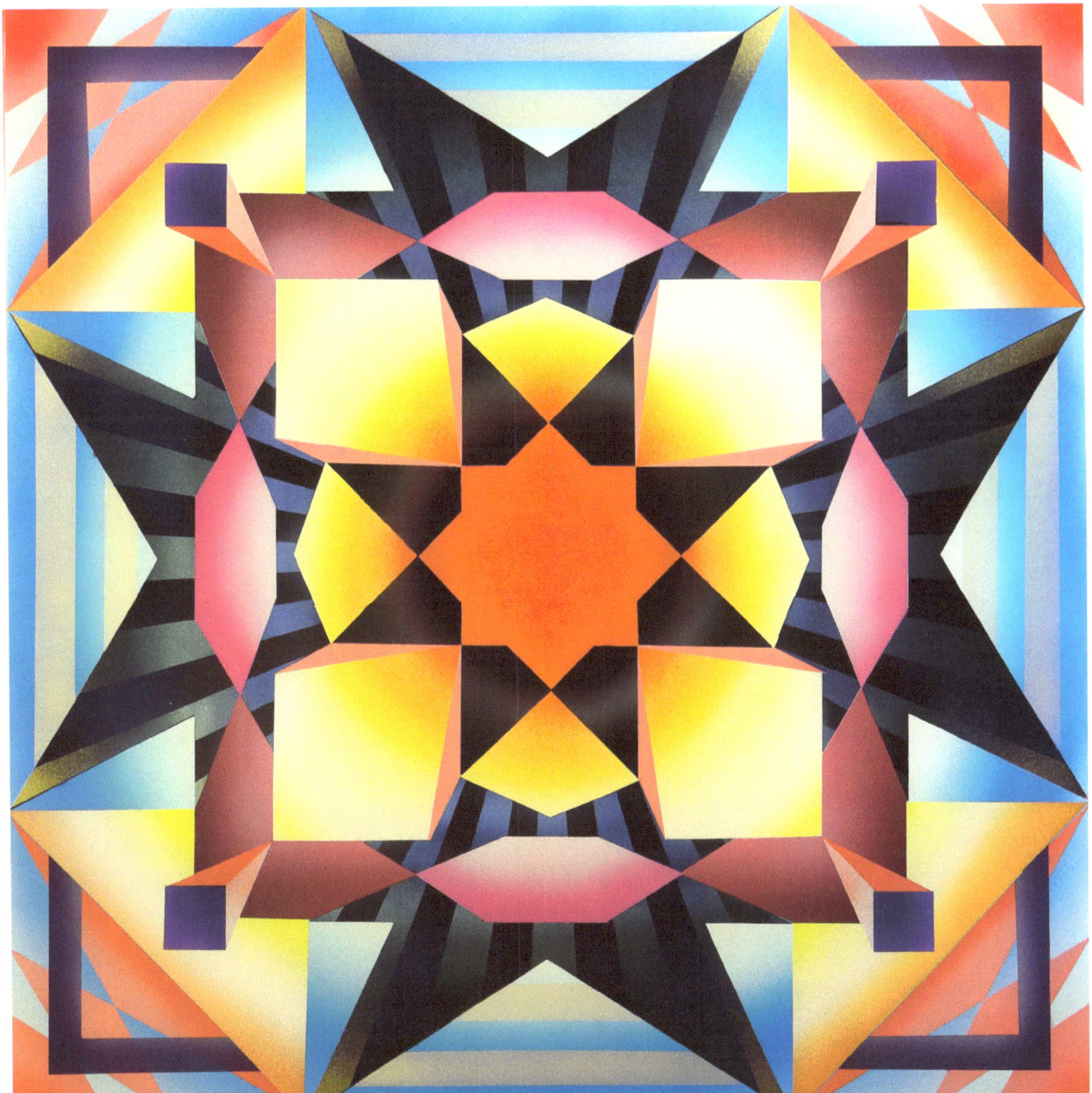

INTRODUCTION

In 2003, I was working at Art Center College of Design in Pasadena, and one of my responsibilities was to serve as a liaison with labs at the California Institute of Technology (Caltech) to find research projects that would be compatible for collaboration with graduate students in the center's Industrial Design program. Discovering the connections and overlaps between the arts and sciences was exciting to me, as was the invitation from Dr. Christof Koch, then a professor at Caltech, to be the first visual artist to attend the Neuromorphic Engineering Workshop in Telluride, Colorado. Eager as I was, I was also nervous; I'd never been to anything like the workshop before, surrounded by brilliant scientists from leading institutions around the world. When I arrived, Mark Blanchard, then a postdoctoral associate at the Institute of Neuroinformatics at the University of Zurich and ETH Zurich, was on hand to help me check in. I sat down next to him at one of the lab computers, and he asked me to create an account by typing in my name and, as a temporary password, "new space user." I typed my name and "new space user." Easy enough, but why did Mark look so confused? "No," he said awkwardly, almost embarrassed. "Type in 'new space user.'" *I did*, I thought, baffled and a bit flustered. I retyped "new space user," only to turn and see Mark's expression more dumbfounded than before. It was only then that I realized what he wanted me to type: "new user," two words, with a *space* between them.

The thought of space as a nothing, an absence—as obvious as Mark's instruction is to me now, it's not surprising that my mind refused to process it at first. In my world, space is more typically a *something*, most often a larger physical dimension populated with material objects (a.k.a. art) and people. Much of my life's work has been making things to occupy space, with space often a reality at once objective and subjective, as implied by the questions that form the basis of so many art-world conversations: "Did you see the space?" "How do you envision the space?" "It's a [blank] space" (insert almost any adjective you wish, from "fabulous" and "fantastic" to "dismal" and "depressing").

As I further explored collaborations with the scientific community in the years after that first conference in Telluride, though, I quickly came to realize that my notions surrounding the concept of space were circumscribed by my own experience. To be exposed to *space* as defined by astrophysics, for example, or psychophysics (the quantitative study of the relationship between external stimuli and sensory perception) felt like a revelation, never more so than in 2018 when I had the privilege of flying on a mission of NASA's Stratospheric Observatory for Infrared Astronomy (SOFIA). Prior to that, Andi Campognone and I had been discussing my participation in *The Light of Space* exhibition at the Lancaster Museum of Art and History (MOAH). Andi has long been a visionary champion of Southern California artists, including myself, and she enthusiastically agreed to a unique collaboration that would be part of my contribution to the show: I would invite a cross-section of scientists, art-world professionals, and others with whom I've crossed paths to contribute essays reflecting on what space means to them, which would then be complimented by Andi's selection of work by contemporary artists who visually address concepts of space in their art.

The result—this book—seeks to explore how our various definitions of space intersect, and how they don't, which in turn becomes fodder for reflection upon our relationship to ourselves, to others, and to the universe at large. It also stands as a testament to the potential inherent in art practices that seek points of connectivity and collaboration outside the realm of the art world, a potential that seems poised to gain even greater visibility with the coming of the next iteration of the Getty's regional arts initiative, Pacific Standard Time: Art x Science x LA, in 2024, which aims to dramatically expand the investigations of the connections between science and art.

Back on Earth after my SOFIA flight, I was struck by yet another revelation, which is that when it comes to (outer) space, there are rules and no rules at the same time. Forward is also backward; up is down; inside is outside; and sideways is also the other way. Confusion is part of the poetry. Location, coordinates, mapping—all of a sudden these occurred to me as elegant and eminently logical, yet also the thinnest sheath of rational order overlying anarchy. Chaotic precision. Balanced turmoil. Thinking about all we've packed into the word *space*, flinging those contents out, and contemplating the beautiful chaos that ensues, the open possibilities—that is what this project is about.

- Shana Mabari
Los Angeles

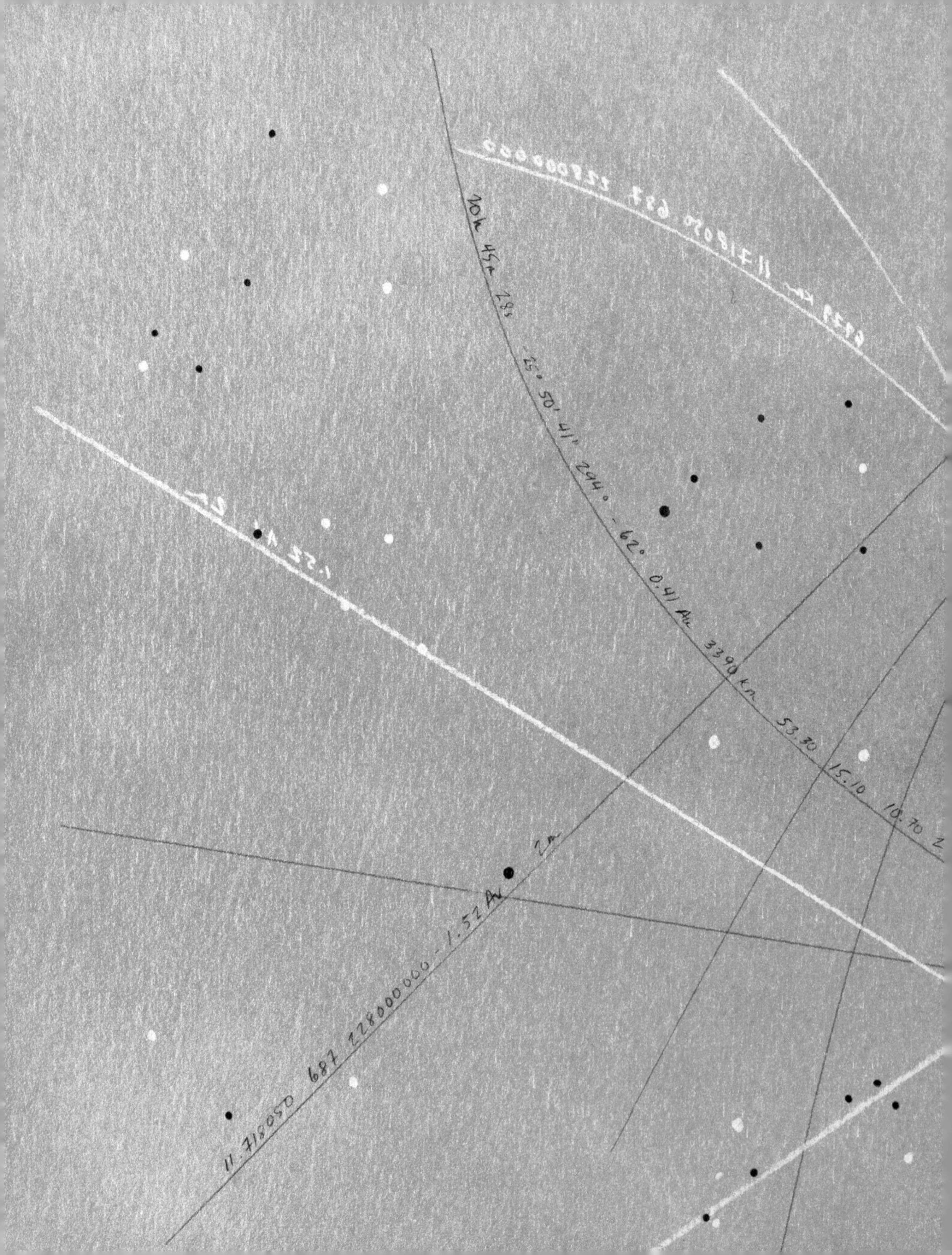
20h 45m 28s -25° 50' 41" 294° -62° 0.41 Au 3390 km 53.30 15.10 10.70 2
11.718050 687 228000000 - 1.52 Au 2m

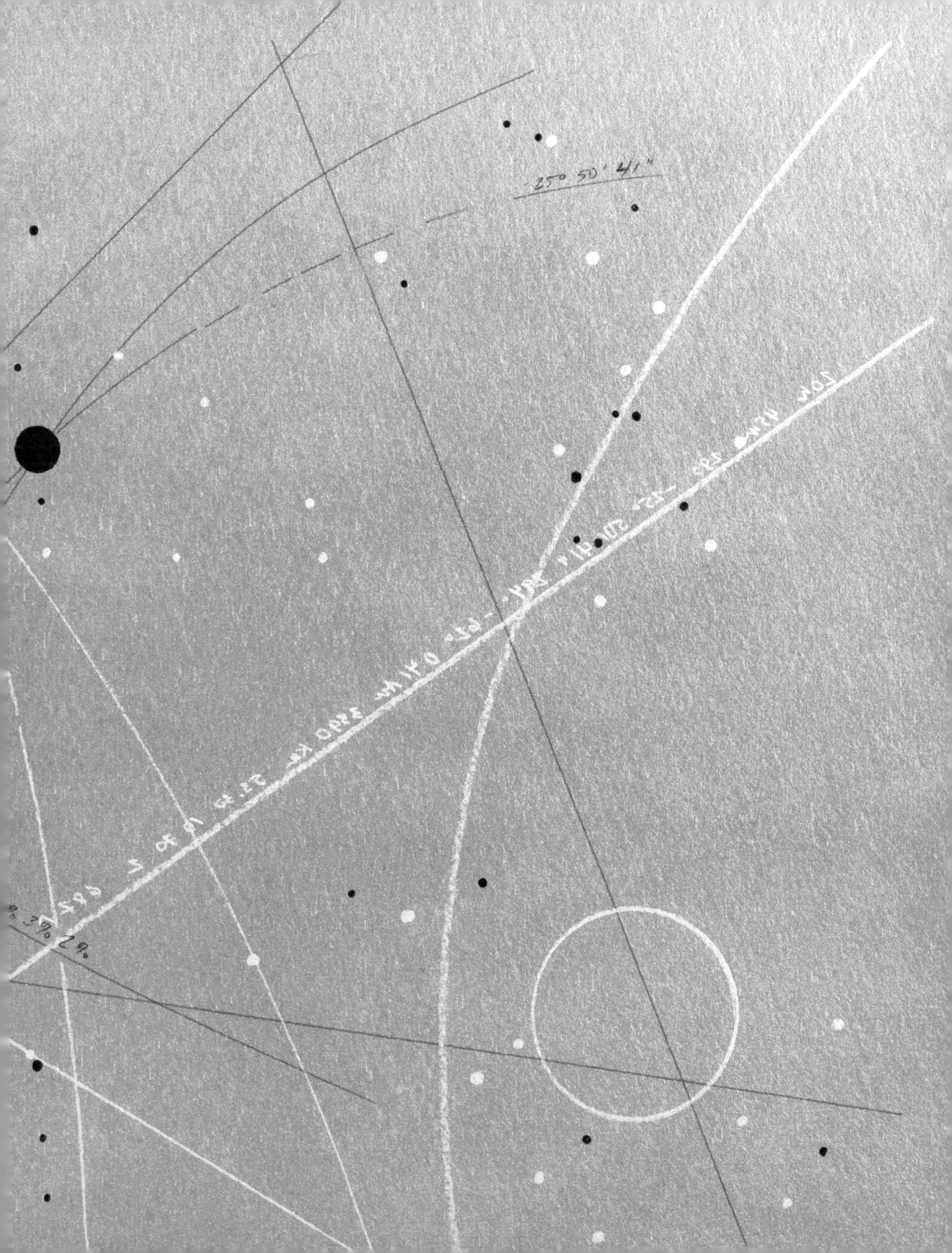
25° 50′ 41″

ACKNOWLEDGMENTS

Any project that aims to represent a multiplicity of voices, almost by definition, relies on the contributions of many individuals, and we are deeply grateful to all those whose enthusiasm and commitment have made this book possible. Our first thanks go to the Lancaster Museum and Public Art Foundation and to publishing house Griffith Moon, whose collaboration and support have been indispensable.

Central to this effort have been the contributions from the artists and writers who responded to our vision for this project with a generosity of time and passion, for which we cannot thank each of you enough. To our essayists, many of whom deserve great credit for their willingness to venture beyond their areas of expertise to partake in creative adventures, we extend our sincerest gratitude for your insightful and engaging original texts: Javier Aparicio, Abbott Brown, Lesley Chilcott, Dr. Kynan Eng, J. Raúl Guzmán, Rami Jaffee, Cooper Johnson, Dr. Janet Kavandi, Dr. Randolf Klein, Dr. Susanna Kohler, Danni Landa, Gary Lang, Shana Mabari, and David McBride. We are most grateful as well to the estate of Leonard Shlain for graciously allowing us to include a quote and excerpt from his book *Art and Physics*.

To the artists who generously granted us the privilege of including their work here, we applaud your commitment to this type of collaborative art making and offer our sincere appreciation for your excellent contributions, which have made this book and its unique combination of images and words an exquisite cocktail of twenty-first-century exotic contemplation: Casper Brindle, Alex Couwenberg, Laddie John Dill, Jeff Frost, Jay

Mark Johnson, Kysa Johnson, Gary Lang, Shana Mabari, Ruth Pastine, Mary Anna Pomonis, Marcia Roberts, Dr. Shinsuke Shimojo, Phillip K. Smith III, Robert Standish, and Edwin Vasquez.

For her part, Shana would like to thank Mark Blanchard for the embarrassing and confusing exchange that ultimately inspired this project, as detailed in her introduction to this book. She extends her deepest appreciation as well to NASA and its partner DLR (German Aerospace Center) for the privilege of flying on board flight #537 of the Stratospheric Observatory for Infrared Astronomy (SOFIA) and for having the vision and foresight to include a visual artist on board such a revolutionary airborne scientific instrument.

We are incredibly thankful for the careful and dedicated work of editor Jason Best and his ability to keep the myriad details of editing aligned with the overarching vision for this project.

Lastly, thank you to the readers who value the potential inherent in a collaborative work such as this one. We hope this book inspires experts in fields that are seemingly a universe apart to find joy and inspiration in the discovery of overlapping interests, complimentary obsessions, and ultimately future collaborations.

- Shana Mabari and Andi Campognone

SPACE AND THE UNIVERSE

In the beginning, space wasn't so big.

At the dawn of time, nearly 14 billion years ago, the infinite possibilities of our universe were compressed into an unfathomably dense, hot point. In the tremendous explosion that marked the birth of the universe, space inflated like a balloon filled by powerful lungs. Scalding-hot particles ricocheted off of one another at incredible speeds, and chaos reigned as matter and antimatter formed and annihilated at random.

Throughout this turmoil, space grew and grew. Eventually, the universe became large enough that it started to quiet. As the temperature dropped, tiny particles began to stick together, forming larger particles; larger particles came together to form atoms. Gravity, unable to halt space's relentless expansion, nevertheless drove the collapse of clouds into the first stars and the organization of stars into galaxies—defiantly building structures even as the space between those structures continued to grow.

Today, our universe is extraordinarily vast, and with each passing second the fabric of space-time stretches and expands even further. As an astronomer, it's sometimes difficult not to become overwhelmed by the immensity of the distances and times involved in what I study—particularly when I consider that while space itself has grown, so, too, has our human ability to perceive its size and depth.

•••

In the lifetime of the universe, humankind's appearance was a heartbeat ago. When humans first arrived, the universe had already spent nearly 14 billion years growing. Earth was just one of many planets dancing around the sun, which was just one of hundreds of billions of stars burning within the Milky Way, which was just one of a trillion galaxies flung throughout the universe. But early astronomers couldn't see this vastness; to them, space was a much smaller place.

Before advances in optics, keen eyes were astronomers' observing tool of choice. Curious gazes could track the motion of the sun or moon as they sailed across the sky, the twinkling of stars painting the dark canvas at night, the steady shine of planets as they slowly wandered. But it was easy to assume that the universe ended at the edge of where the human eye could see.

It wasn't until the seventeenth century that a new discovery opened up man's ability to perceive the vast scales of space, a device that used pieces of curved glass to gather light from distant heavenly bodies and make them appear larger. With Galileo's first use of a telescope for astronomy, space began to open. The telescope revealed the phases of Venus, evidence that Earth was not the center of the universe. It revealed the moons of Jupiter, proving that worlds upon worlds lay just beyond our own. It revealed that the bright, milky cloud overhead was not a nebula, as believed, but a dense collection of tiny pinpricks of light—an impossible number of distant stars spilling across our night sky.

As technology advanced, our tools for observing the universe grew more powerful. We explored tails of icy comets that streaked across the sky, marveled at mysterious clouds of gas and dust, and measured the distances to stars. New observations in the last century showed us that fuzzy spiral nebulae were actually their own "island universes"; our Milky Way was just one of many galaxies. Quasars—blazing cores of enormous galaxies—were spotted billions of light-years away. Each new discovery drove home the same point: space is larger than we'd ever imagined.

By the time I entered the field of astronomy more than a decade ago, not only had our perceptions of space's size grown, but also of its richness. Over the course of my career, we've discovered thousands of worlds—many just like Earth—orbiting stars of all flavors. We've learned about the roiling interiors of stars and peered deep into the quiet, empty spaces between them. We've observed delicate planetary rings woven from gossamer threads of gas and dust, and we've watched black holes collide, creating ripples that shake the very fabric of space-time. We've even witnessed the faint background glow from the universe's initial explosion—fingerprints from the dawn of time.

•••

In the beginning, space wasn't so big. But what will it be in the end?

For all our observatories and all our theories, astronomers aren't yet certain of our universe's fate. Staring up at the inky abyss on a clear night, I often contemplate the likely outcome: if space continues its unrelenting expansion, we march ever into a colder and darker future. As the space between galaxies increases, it takes longer and longer for the light from other worlds to reach us, and that light becomes dimmer and dimmer. Eventually we will be alone in the dark.

I contemplate this prospect without sadness; indeed, it only enhances the beauty of the twinkling paint splashed across the canvas of the sky. If the future is dark, then how fortunate are we—as brief inhabitants of this rock, suspended in the vastness of space—to be here in this time when light is plentiful and discoveries are bountiful? How fortunate that we have made such progress in understanding the tremendous scales of our universe and the diversity of objects hidden within the night sky's tantalizing depths. And how fortunate that there are still so many mysteries left to uncover and puzzles left to delight over, out here in the midst of expanding space.

- Dr. Susanna Kohler
Editor, AAS Nova
American Astronomical Society

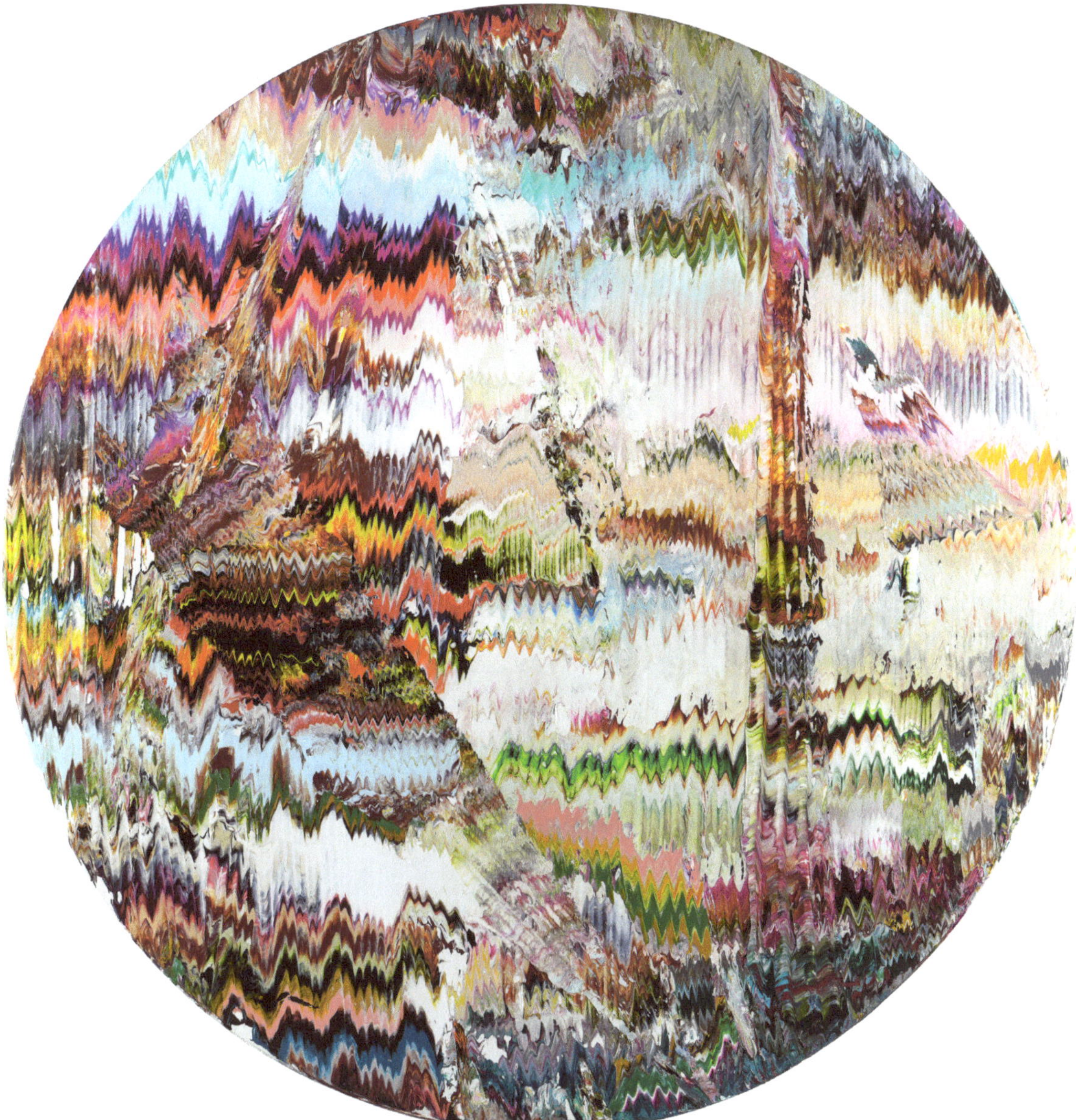

SPACE AND WORDS

still   hushed   silent
    the magic of the tongue;
intrigue of the mind,
    maddening effects of loneliness, and
the allure of 'Breaking News'
    is remote

liberated from seams and contours of intention,
    the mediocrity of fear, and
the fervent prayers of pilgrims in paradise
    one million Earths fit inside an average star

following the ecliptic
    planets file in formation
all versions of hell,
    the idleness of want, and
cheerful grief
    on pause

the world ceases to hold its breath
    the demonic recedes from the hearts of men
we dwell in possibilities,
    rather than the tragedies
of habitual luxury, and
    the absurdity of a good fence

warps, ripples, and
    distortions in the fabric of time
fleeting subatomic particles force space apart
    horizontal diamond and glass hailstorms,
NVIDIA's rise, and
    streams of atoms play G-D,
silencing the heroics of ambition,
    organized indifference, and
the physics of 'what if'
    uniting the ONE

zoos of pulsars, spikes of heat, and
    nodules of trapped energy
devoid of the pretense of meaning, and
    the gravity of relevance
enigmatic disruptions with gravity,
    matter and antimatter collide and annihilate each other
converting mass into pure energy
    septillions of stars neither static nor timeless
collapse and fold, connecting routines, patterns, and
    coincidence in the vastness between words, deeds, and
a 'peace worth fighting for'

    from a distance of 165 quadrillion miles,
an insignificant planet
    tucked away in a corner of the universe
teaming encampments of weary souls, and
    heartless narcissists gather G-D's gold,
accumulate error, nibble at the margins of trembling insubstantiality, and
    cluster at the last landfill pondering meaning

simultaneously bioluminescent inhabitants couple by chance
    entering the heart of the world, where
we breathe the perfume of all things
    filling space with cascades of kindness and
tranquil strength

a combat of lightning bolts
subdued by erotic current and
genital fire, populate infinitudes
braving fortune in rhumba heaven
delirium on the frontiers of being
daring to embrace forever . . .

2.468542065e+23 trillion kilometers
26092510215.81 light-years,
8 billion parsecs . . . alas
the determination of the quiet ones
rogues of solitude favoring chance;
interrogating vision, juggling the sublime

cavemen whirling in a primal ocean,
channeling radiance with ever-present heart
mirroring the hieroglyphics of ecstasy,
acknowledging the indelible mystery and
thunderous roar of uncompromising love
imprinting Infinity

- Gary Lang
Artist

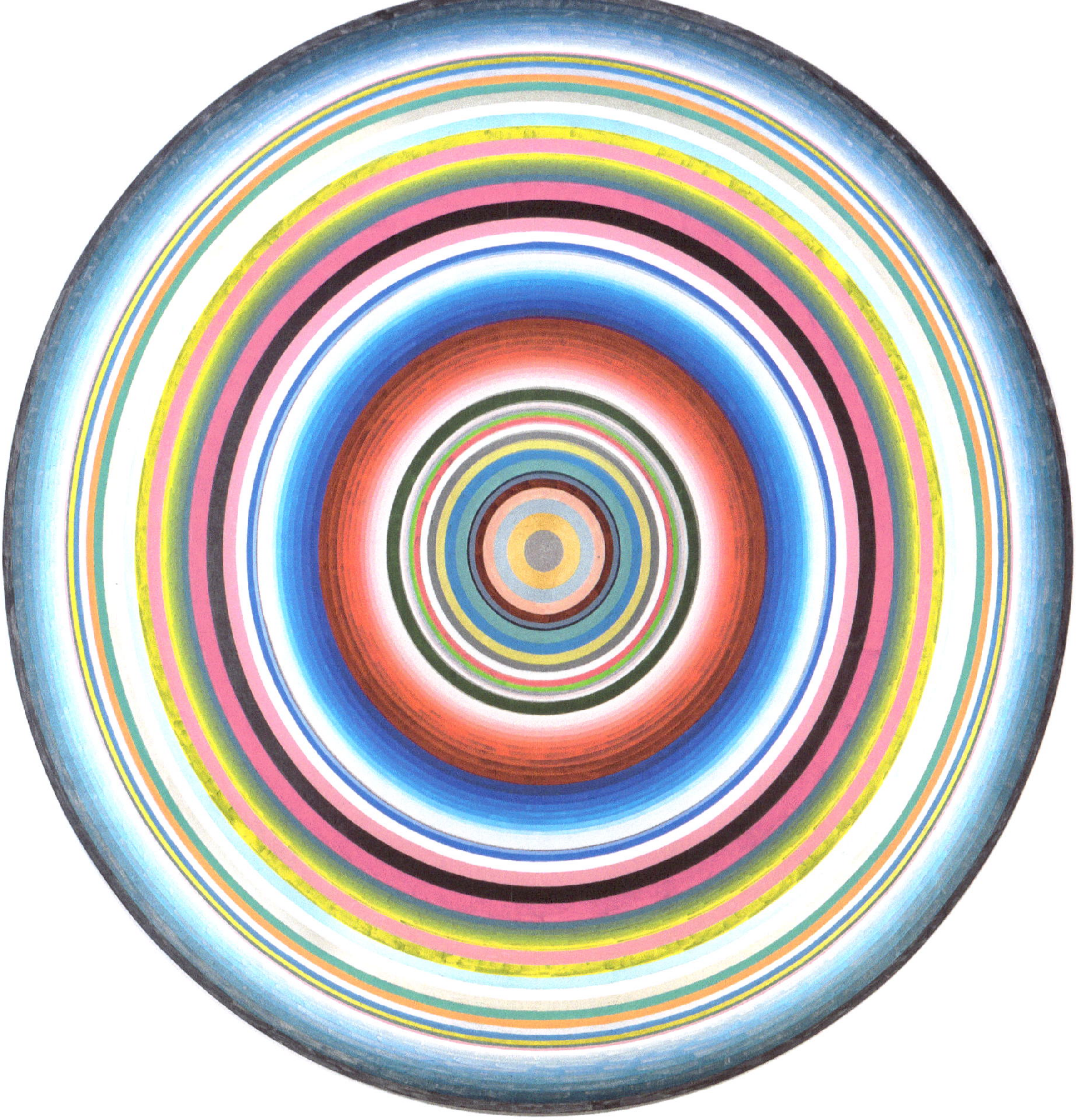

RAUM UND ASTRONOMIE

"Space, the final frontier . . ."

Das ist das Erste was mir zu dem englischen Wort 'space' einfällt, was nicht weiter verwunderlich sein sollte, denn viele Astronomen sind von 'Raumschiff Enterprise' beeinflusst worden. Da ich in Deutschland aufgewachsen bin, lauteten die Eingangsworte allerdings "*Der Weltraum. Unendliche Weiten . . .*". Das mag der Grund dafür sein, dass ich mit dem englischen Wort „space" zuerst den Weltraum mit seinen unendlichen Weiten assoziiere und weniger andere Bedeutungen wie Parkplatz (parking space), Intimsphäre (personal space), oder die mathematische Bedeutung wie in Vektorraum (vector space).

Die enormen Ausmaße des Weltraums faszinieren mich. Trotz der unvorstellbaren Entfernungen sind wir in der Lage zu verstehen, was weit von der Erde passiert, zum Beispiel wie Sterne im Orion Nebel entstehen, oder dass im Zentrum unserer Galaxie ein super massereiches Schwarzes Loch vorhanden ist, oder was passiert wenn Galaxien kollidieren, um nur drei der nahezu unzähligen astrophysikalischen Themen zu nennen.

Wenn ich gelegentlich in Schulen oder bei anderen Gelegenheiten Öffentlichkeitsarbeit mache, dann sind die unvorstellbaren Entfernungen, die in der Astronomie auftreten, eines meiner Lieblingsthemen. Die enorme Größe des Sonnensystems ist schon schwer zu begreifen. Man kann versuchen die relativen Größen der Himmelskörper in unserem Sonnensystem mit maßstäblichen Modellen zu erläutern, wie zum Beispiel dem Pfefferkorn-model. Wenn wir die Erde auf die Größe eines Pfefferkorns schrumpfen, schrumpfen im gleichen Maßstab die Sonne auf die Größe einer Bowling Kugel und der Jupiter zu einer Kastanie. Während man sich das vielleicht noch vorstellen kann, sind die Entfernungen, auf gleiche Weise skaliert, erstaunlich. Die Pfefferkorn-Erde müsste 24 Meter von der Bowlingkugel-Sonne entfernt platziert werden und der Kastanien-Jupiter 123 Meter entfernt. Der äußere Rand des Sonnensystems wäre mehr als 1000 Meter entfernt. Und in diesem riesigen Volumen, einer Kugel mit einem Radius von 1000 Metern, sind die einzigen Objekte eine Bowlingkugel, eine Kastanie, ein Pfefferkorn und all die anderen Planeten, Asteroiden etc. zusammen genommen bilden gerade mal eine zweite Kastanie. Und all dieser leere Raum ist nur unser Vorgarten. Der nächste Stern, eine weitere Bowlingkugel, müsste in einem anderen Kontinent, 6500 Kilometer entfernt, platziert werden. Und da wäre nichts als leerer Raum zwischen den beiden Bowlingkugeln. Dieses 'Erde als Pfefferkorn' Model lässt vielleicht den Raum erahnen den unser Sonnensystem einnimmt, vielleicht sogar die Entfernung zum nächsten Stern aber das ist nichts im Verhältnis zur Größe unserer Galaxie, der Milchstraße. Wenn wir unser Sonnensystem auf die Größe einer 1€ Münze schrumpfen, da wäre unsere Milchstraße so groß wie die Vereinigten Staaten. Wir könnten so weiter machen und unsere Milchstraße auf die Größe einer 1€ Münze schrumpfen und uns fragen, wie weit entfernt die nächste Galaxie in diesem Maßstab wäre (60cm wären es zur Andromeda Galaxie, der nächsten Galaxie ähnlicher Größe), aber die Größe unseres ersten Models des Sonnensystems ist schon unvorstellbar klein geworden. Der Weltraum, die Größe von Sternen und Galaxien und deren Entfernungen untereinander entschwinden sehr bald unserer Vorstellungskraft und hinterlassen ein Gefühl wundersamen Erstaunens.

Und trotz der unvorstellbaren Ausdehnung des Weltraumes können wir ihn doch mit Hilfe von Zahlen und Mathematik begreifen. Die ist eine elegante und erfolgreiche Kombination. Die Unendlichkeit des Zahlenraumes ist eine gute Analogie zum Weltraum und all die Werkzeuge der Mathematik ermöglichen es uns astronomische Beobachtungen zu verstehen und Entfernungen und Abläufe im Weltall zu begreifen. Das ist nicht nur der Fall, wenn wir den Weltraum und astrophysikalische Phänomene verstehen wollen. Die Mathematik ist die Sprache aller Naturwissenschaften. Der Erfolg dieser Sprache in der Beschreibung der Welt ist erstaunlich. Die

Entsprechung zwischen dem Reich, oder Raum, der Mathematik und dem Universum ist elegant, mysteriös und faszinierend. Der Nobelpreisträger Eugene Wigner schrieb dazu: „Das Wunder der Anwendbarkeit der Sprache der Mathematik auf der Formulierung der Gesetze der Physik ist ein wundervolles Geschenk, dass wir weder verstehen noch verdienen."

- Dr. Randolf Klein
Science Outreach Lead
SOFIA Science Operations, NASA

1. Guy Ottewell, "The Thousand-Yard Model, or the Earth as a Peppercorn," National Optical Astronomy Observatory: https://www.noao.edu/education/peppercorn/pcmain.html.

2. "How Big Is Our Universe? An Exploration Through Space and Time," Universe Forum, Harvard Smithsonian Center for Astrophysics: https://www.cfa.harvard.edu/seuforum/howfar/across.html.

3. Mit der Erwähnung der „Unendlichkeit des Zahlenraumes" möchte ich nicht sagen, dass der Weltraum unendlich ist. Wir wissen nicht, ob er endlich oder unendlich ist.

4. Eugene Wigner, „The unreasonable effectiveness of mathematics in the natural sciences. Richard Courant lecture in mathematical sciences delivered at New York University, May 11, 1959," *Communications on Pure and Applied Mathematics* 13, no. 1 (February 1960).

SPACE AND ASTRONOMY

"Space, the final frontier . . ."

That's the first thing that comes to my mind when I think about the word *space*, which shouldn't be a surprise, really—many astronomers have been influenced by Star Trek. Because I grew up in Germany, though, the iconic opening words for me were "*Der Weltraum. Unendliche Weiten ...*," with *Weltraum* (literally "world space") specifically meaning outer space or the universe, and *Unendliche Weiten* translated as "infinite expanse." This may be the reason why my primary association with the word *space* is the universe in its infinite expanse, much more so than its other meanings, like in parking space, personal space, or its mathematical meaning as in vector space.

The vastness of space is what fascinates me. And yet still we are able to learn what happens at unimaginable distances away from Earth, such as how stars form in the Orion Nebulae, or that there is a supermassive black hole in the center of our galaxy, or what happens when galaxies collide, to mention only three of the (seemingly infinite) topics of astrophysics.

When I do occasional public outreach work in schools or elsewhere, my favorite activities revolve around trying to explain the mind-boggling distances that are involved when talking about space and the sizes of astronomical objects. The enormity of our solar system alone is difficult to grasp. One can try to explain the relative sizes of the bodies in our solar system by making a scale model, like the peppercorn model.[1] If we scale Earth down to the size of a peppercorn, the Sun scales to a bowling ball and Jupiter scales to a chestnut. While this may be relatable, the distances scaled the same way are astonishing. The peppercorn Earth needs to be 26 yards away from the bowling ball Sun; the chestnut Jupiter ends up 135 yards away from the bowling ball Sun. The outer edge of the solar system in our scale version would be more than a thousand yards away. And in all this vast space—a sphere a thousand yards in radius—the only objects are a bowling ball, a chestnut, a peppercorn, and a relative smattering of planets, asteroids, and other things that taken together would form but a second chestnut. All this empty space, however, is just our back yard. The next closest star (another bowling ball) needs to be placed on an entirely separate continent *four thousand miles* away, with nothing but empty space between the two bowling balls. This peppercorn Earth scale model may allow us to grasp the space taken up by our solar system, maybe even the distance to the next star, but that is nothing compared to the scale of our galaxy. If we scale our solar system down to a size of a quarter, for example, our galaxy the Milky Way would be the size of the United States.[2] We could go on, devising yet another scale model where the Milky Way is the size of quarter to explain how far away the next galaxy would be in that scale model (two feet to the Andromeda galaxy, the closest galaxy of similar size). But then the size of our first scale model of our solar system quickly starts to become unimaginably small. Space, the size of stars and galaxies, and their distances relative to one another, soon slip again beyond our comprehension, leaving us with the thrill of wonder.

Despite the unimaginable vastness of space, we can still make sense of it with the help of numbers and mathematics. This a beautiful and very fruitful combination. The infinite space of numbers is a good analogy to space, and all the tools that mathematics provides allow us to work out what our astronomical observations mean, what they tell us about the distances and processes in space.[3] This is true not only when we try to understand space and astrophysical phenomena—mathematics is the language of the natural sciences. The success of this language in describing the natural world is astounding. The correspondence between the realm—space, if you will—of mathematics and the universe is beautiful, mysterious, and fascinating. As Nobel laureate Eugene Wigner once wrote: "The miracle of the appropriateness of the language of mathematics for the formulation of the laws of physics is a wonderful gift which we neither understand nor deserve."[4]

- Dr. Randolf Klein
Science Outreach Lead
SOFIA Science Operations, NASA

1. Guy Ottewell, "The Thousand-Yard Model, or the Earth as a Peppercorn," National Optical Astronomy Observatory: https://www.noao.edu/education/peppercorn/pcmain.html.
2. "How Big Is Our Universe? An Exploration Through Space and Time," Universe Forum, Harvard Smithsonian Center for Astrophysics: https://www.cfa.harvard.edu/seuforum/howfar/across.html.
3. By referencing the "infinite space of numbers," I am not saying that the universe is infinite. We don't know whether the universe is finite or infinite.

4. Eugene Wigner, "The unreasonable effectiveness of mathematics in the natural sciences. Richard Courant lecture in mathematical sciences delivered at New York University, May 11, 1959," *Communications on Pure and Applied Mathematics* 13, no. 1 (February 1960).

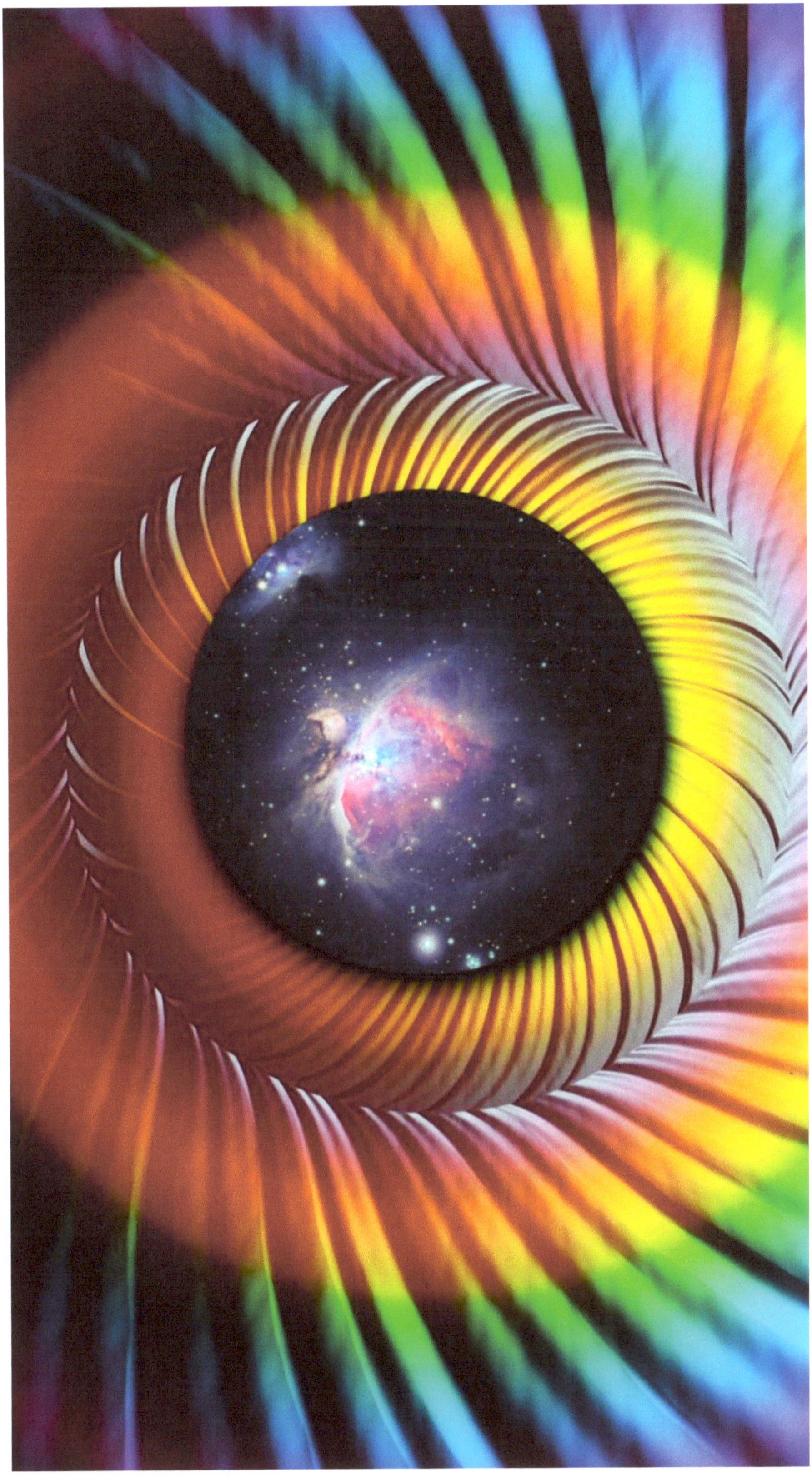

SPACE AND EXPLORATION

When I was growing up, the word *space* was generally followed by "the final frontier." *Start Trek* was the iconic television show, and everyone knew the catch phrase. I suppose it's not surprising then that I have always associated *space* with space exploration.

I was raised on a farm in the Midwest. On warm summer evenings, my father and I would sit outside to enjoy a variety of light shows. At dusk, the fireflies would appear, making a beautiful show with their tiny beacons. My sister and I would run and chase the little flashing drones until it was too dark to see. I would then join my father on the back porch to watch the really big light show. As we lay back on our reclining chairs, we never failed to be awed by the incredible display of stars and the way that the Milky Way stretched across the sky. Almost without fail, we would be rewarded by an occasional shooting star, and I would frantically try to make a wish before it disappeared. I never succeeded.

The next big reward was to spot a smoothly gliding point of light crossing from one horizon to the other. It was not a shooting star but an object sent to space by humans. My dad told me that some of those gliding points of light even had people in them, astronauts who traveled through the heavens in spaceships. At that point in time, I had no idea how a real-life spaceship worked, nor how on earth anyone got to be an astronaut. I just knew that I wanted to be inside one of those points of light and to look back at my house from space.

As I grew older, more and more people flew into space. Some even went to the moon. I remember being at my grandmother's house and waiting for Neil Armstrong and Buzz Aldrin to step out of their lunar lander and take the first steps onto the moon. It was getting late, and my grandmother wanted us to go to bed. She said it wasn't real after all—they were just pretending. But I knew better. When she was born, there were no cars, much less airplanes or spaceships. The thought of traveling through space to the moon seemed too incredible to be true. Despite my wild objections, I remember her walking over and turning off the television, leaving me dumbfounded and incredibly disappointed. I later met Neil Armstrong and Buzz Aldrin, but I never had the nerve to tell them that I didn't watch their first steps on the moon live because it was past my bedtime.

I did everything I could think of to qualify myself for the Astronaut Corps. I went to college and went to college some more, after which I worked for an aerospace company to get as much experience as I could. I applied to NASA every year, and every year I was rejected—until the year I wasn't. I was finally invited to interview, and I somehow made the cut.

A few years later, there I was, sitting in that spaceship on the launchpad, anticipating leaving the surface of the planet. As the engines roared and we lifted off the pad, I cried, not from fear but from the sheer exhilaration and joy of finally achieving the freedom to escape the gravity of Earth and finding that place I had sought from childhood. We soared faster and faster until it was difficult to breathe, and we went faster still. After eight and a half minutes, the main engines cut off. I removed my harness and floated out of my seat. It was amazing! I quickly reached the flight deck and looked out the window for the first time. There below me was our beautiful planet and the homeland from which I had looked up so many years before.

The mission was very busy, but we tried to take a moment to look out the window every chance we got. Most of the time we looked at Earth, but sometimes we were able to look into deep space. I was able to see constellations I had never seen before. The stars were so numerous that the density of the tiny sparkling dots was almost incomprehensible.

The night before we deorbited, we set aside a full ninety minutes so that we could circumnavigate the entire planet without interruption. We watched the United States in daylight and tried to pick out our hometowns. Then we passed into darkness and were entranced by the moonlight reflecting off of the oceans. We watched the lights of all the cities in Europe appear at night as we silently glided overhead. We flew over London and then Paris, and we saw Rome in that perfectly outlined boot in the Mediterranean. Then the Nile Delta came into view, lit up like a giant upside-down Y as we approached from the north. As we entered the relative darkness above the African desert, massive thunderstorms lit up the sky below, like popcorn popping underneath us. We even saw a shooting star, only now it passed *beneath* us as it burned up in Earth's atmosphere. I still wasn't fast enough to make a wish, but it didn't matter so much anymore, as I was living my dream at that very moment.

And then it occurred to me that I was now myself a pinpoint of light streaking across the night sky, like the ones I had watched with my father on our Missouri farm. Perhaps my spaceship could serve as the beacon for other little girls and boys who might want to seek out the escape of gravity and to experience for themselves the incredible view of our home planet. They could follow my lead as I followed the leads of Neil Armstrong and Buzz Aldrin—and yes, even Captain Kirk—into space, the final frontier.

- Dr. Janet Kavandi
NASA Astronaut and Director
NASA Glenn Research Center

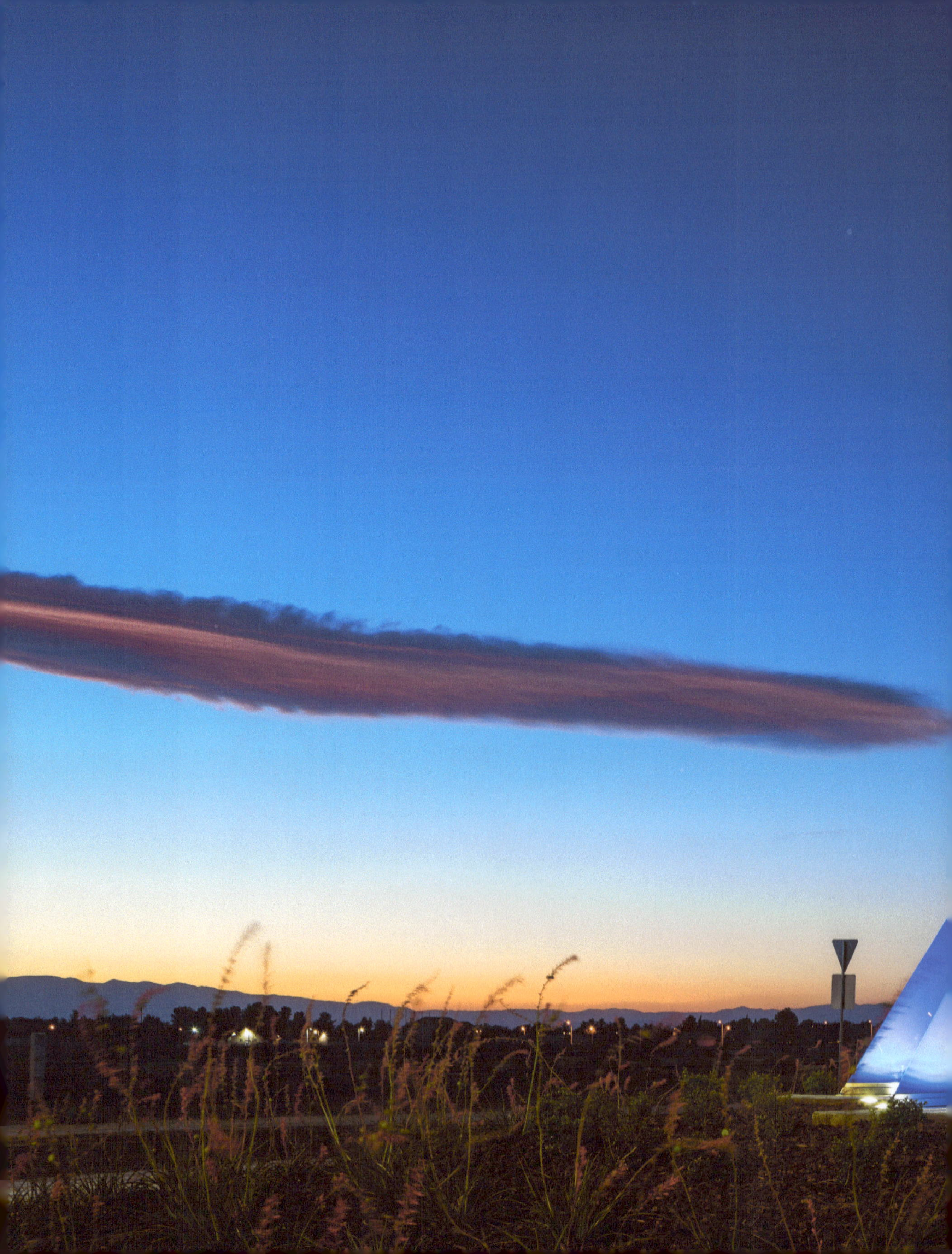

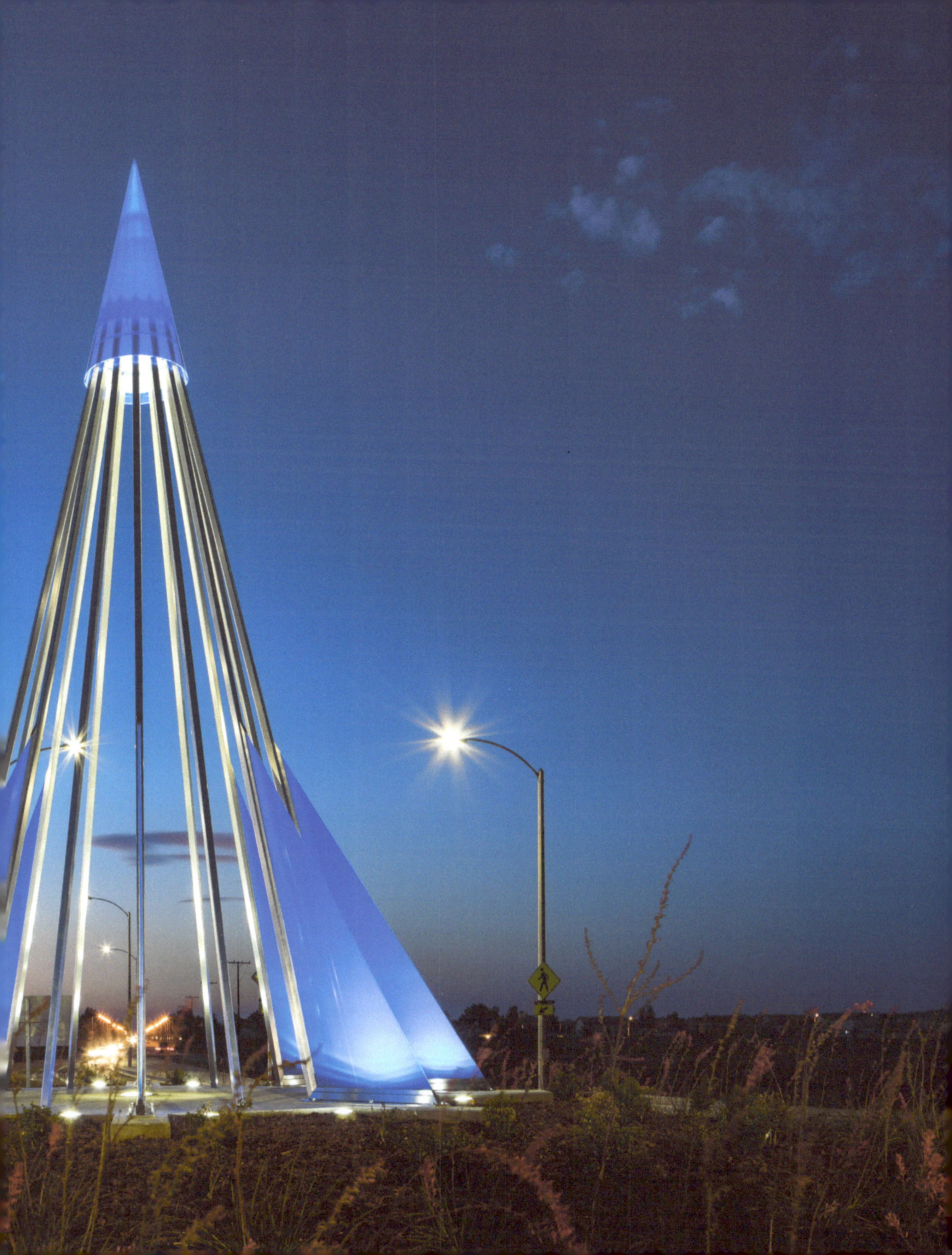

SPACE AND MUSIC

Sometimes my favorite notes and beats are the ones *not* played. That's what I think of when I think of space in relation to music. Time and again in the studio I hear myself calling out to other musicians, "I really love what you *didn't* play," which sometimes gets a double take, like "huh?" But what I'm trying to do is to get more people to "hear" what's not played and to recognize the importance of that kind of space in music. It takes a certain kind of genius to *not* play something. More than ever, especially in pop music, today's audiences are attuned to what you might call "less is more" music, but ironically that can cause a crisis for recording musicians and producers trying to appease those audiences because even in simple, pared-down combinations, there are so many notes to choose from. Soon, "less" becomes more, and then more and more and more! That's why it's so important to cultivate the genius that is understanding what not to play, to bring that space into music more, and to realize that what's not played is as essential as what is played, kind of like how in visual arts, negative space works with positive space to create depth and volume.

I've often thought what would it be like to have an award for what a musician didn't play? That might sound strange, but the importance of space in music deserves more recognition than sound sometimes.

- Rami Jaffee
Keyboardist
Foo Fighters and The Wallflowers

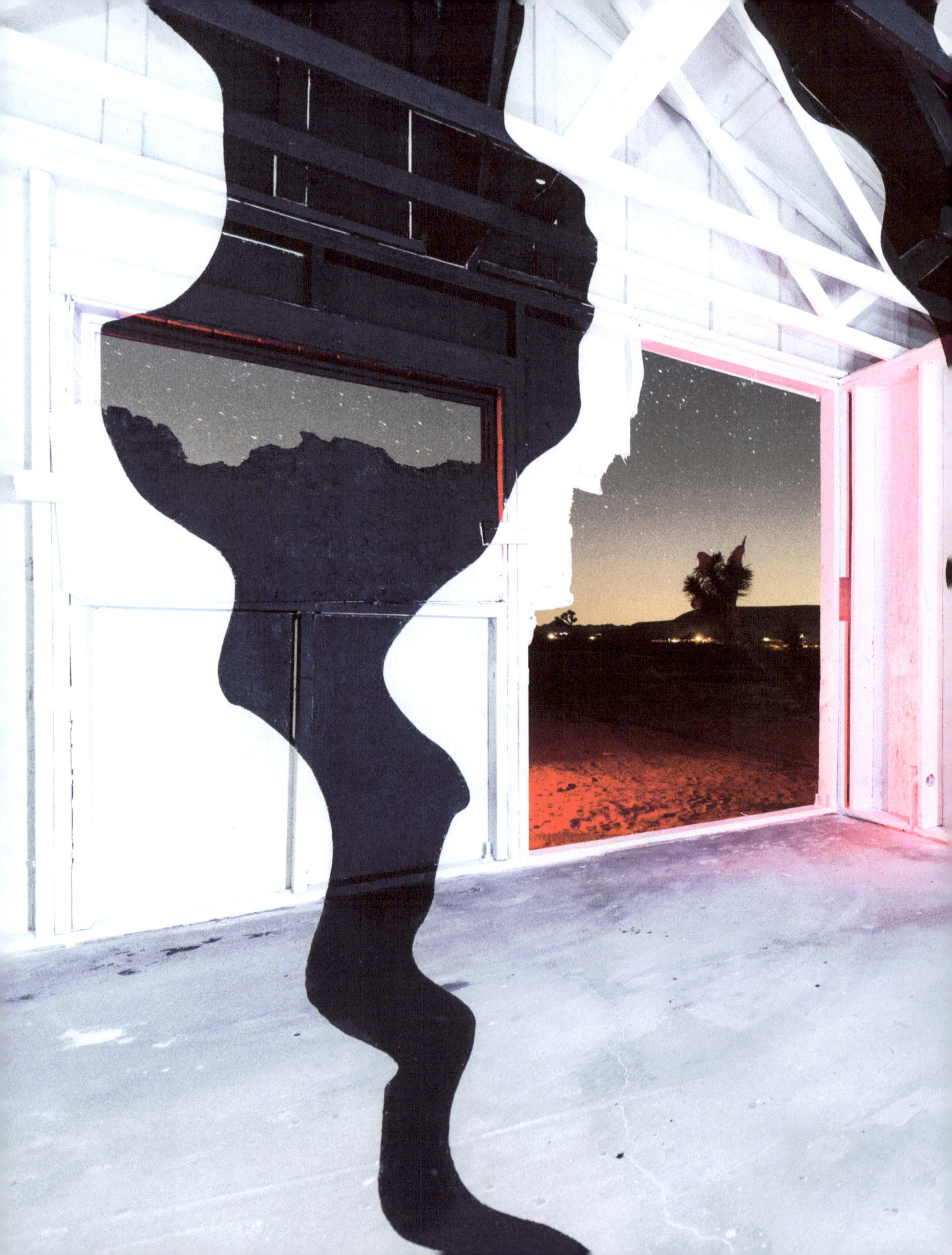

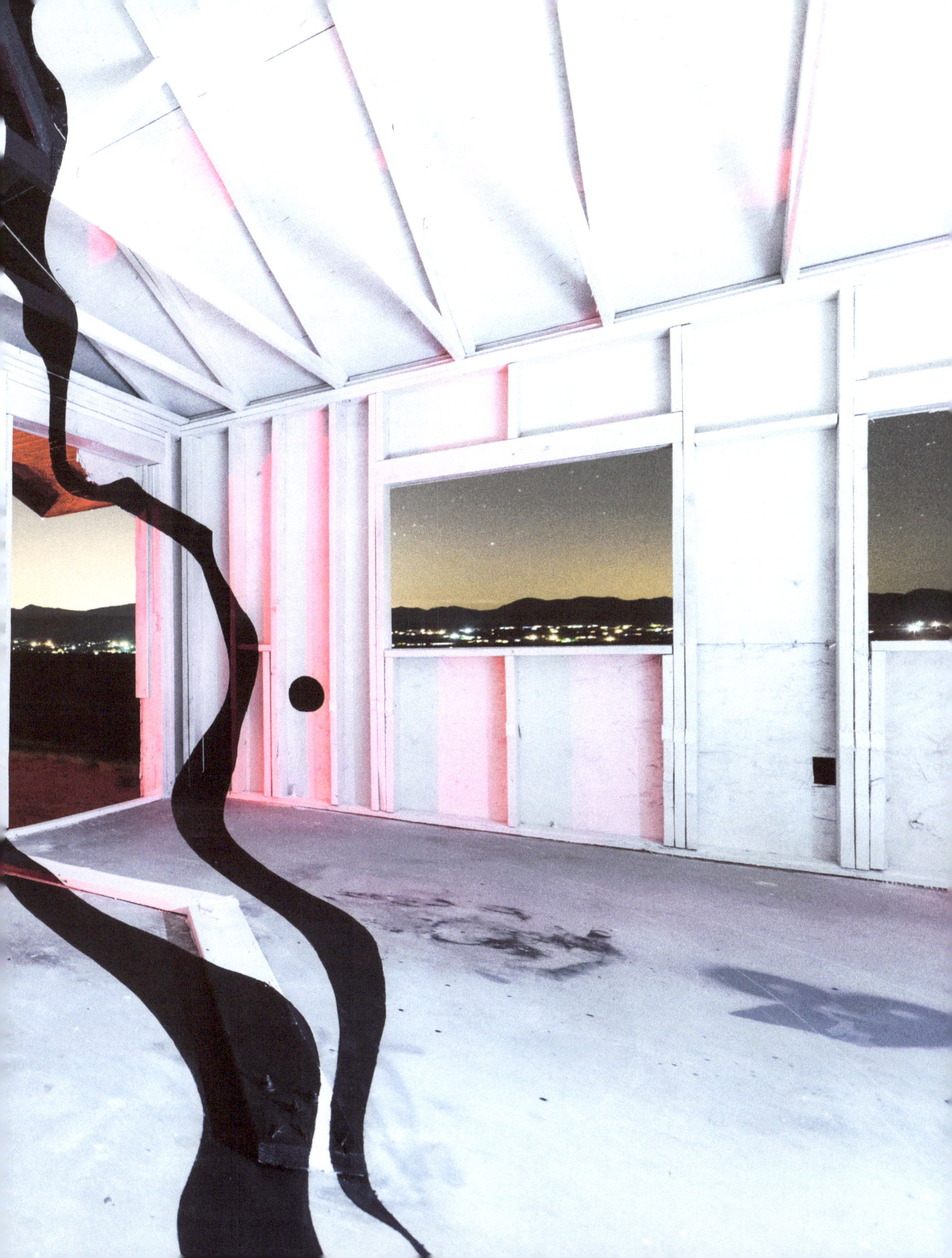

SPACE AND . . . SPACE

The word *space* appears in the official name of NASA, and there is no doubt about what it means in the context of an agency that was established in the late 1950s, as the "Space Race" between the United States and Soviet Union was kicking into gear. With breathtaking speed, fueled in equal measure by fear and by faith in technological progress, the two superpowers swiftly dominated a realm that for much of the late nineteenth and earlier twentieth centuries had been the sole province of sci-fi fantasy, and before that a black void pocked with stars made to reflect the myths and religious mysteries of earthbound humans. The frontier of this space—indeed the *space* in the name National Aeronautics and *Space* Administration—is typically thought to be the lowest altitude at which Earth satellites can stay in orbit without falling into the Earth's atmosphere, defined as approximately 100 kilometers, or 62 miles, above sea level, though in reality the line varies somewhat depending on a variety of factors. This is the space occupied today by communication satellites, the GPS constellation, and the International Space Station, human-tended continuously since November 2000. It is the space occupied by the moon, the planets, the sun, the stars, and galaxies in the cosmos out to the edge of the universe. It is the space that continues to fuel our imaginations and our desire to explore, as attested by NASA's recently inaugurated Artemis program, which will return humans to the Moon and, ultimately, send them on to Mars.

Yet there is another space implied in NASA's name: aeronautical space. Technically it's the space from (almost) zero altitude—a molecule above Earth's surface—to that invisible demarcation about a hundred kilometers up. This is the space we occupy, the space we move in. It is the space most useful to us, the space that provides the fundamentals of life. It is the space we take for granted but also the space just beyond our arm's reach that we set our sights on long before we dreamed of reaching for outer space. This aeronautical space is the space in which we learned and conquered flight through the simplicity of buoyance, the elegance of aerodynamic lift, and the brute force of ballistics. We have become ever more reliant on this space: for our economies, our defense, our leisure, and our understanding of Earth and the cosmos beyond.

We move people, we move material, we move dreams in this space.

We protect, we seek understanding and knowledge in this space.

We understand better the health of Earth from this space. We observe astronomical phenomena from this space, from the planets in our solar system to the nuclear engines of distant galaxies.

We have needs that utilize this space and, to an ever-increasing extent, needs that utilize the space beyond—outer space. Humankind has dreamed of flight since before Icarus, and for centuries now we have been imagining, engineering, and building flying contraptions to meet all those needs as we continue to strive to understand and improve the mechanics of flight.

We think of the process and mechanics of flight in this space. We bring flying contraptions into being to operate in this space. In developing these flying contraptions, our tendency is first to think of engineers and technicians. But flight in both aeronautical and astronautical space requires design, the vision of artists and architects; visual arts professionals have contributed much to both aero- and astronautical flight. From the elegant curve of the hot-air balloon to the iconic shape of a 747, the sleek aerodynamics of rockets to futuristic visions for spacecraft that might ferry ordinary citizens into outer space, the synthesis of engineering and visual artistry in the service of flight has yielded forms that captivate our imaginations and continue to inspire our most far-reaching ambitions in both aeronautical and astronautical space.

- David McBride
Director
NASA Armstrong Flight Research Center

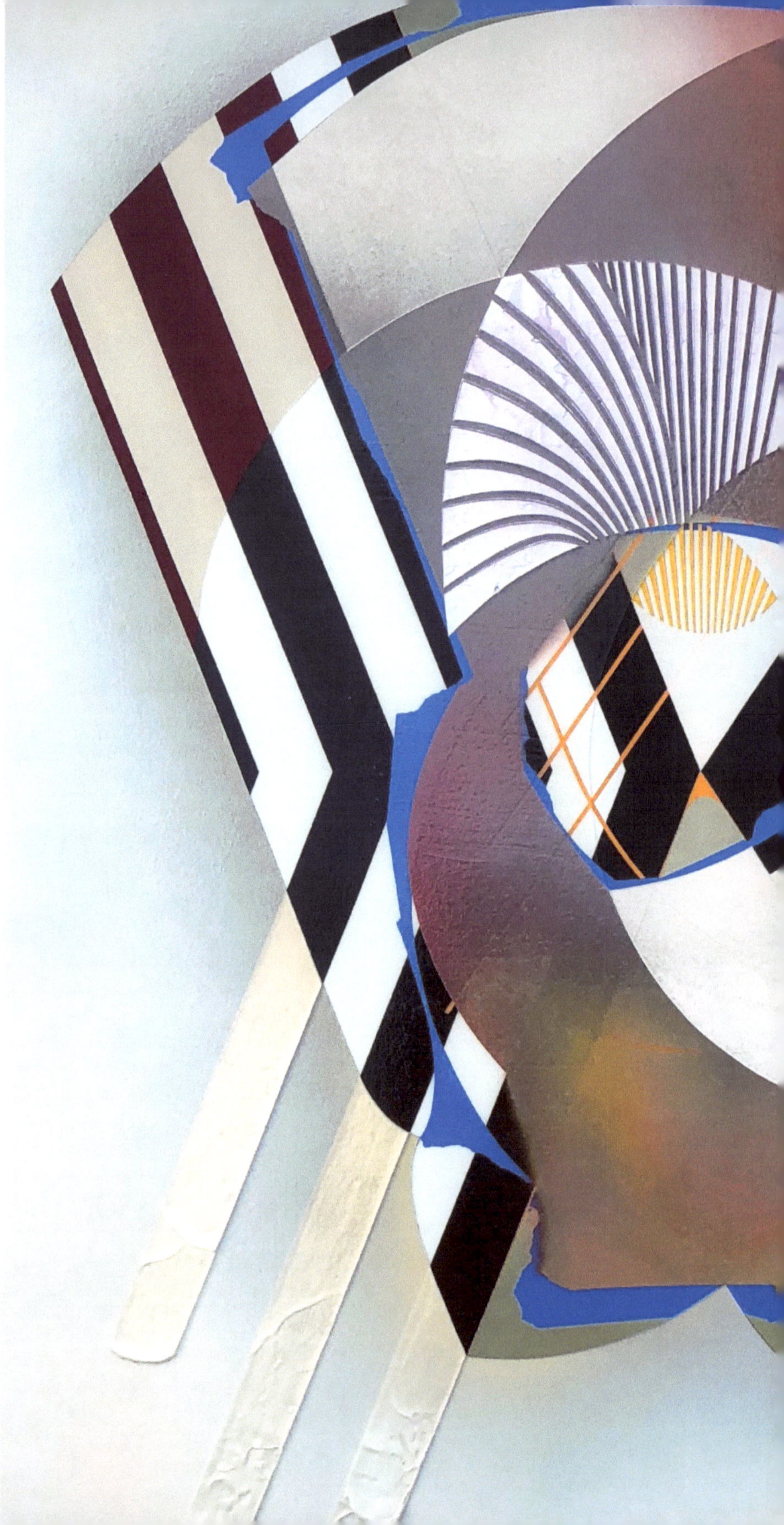

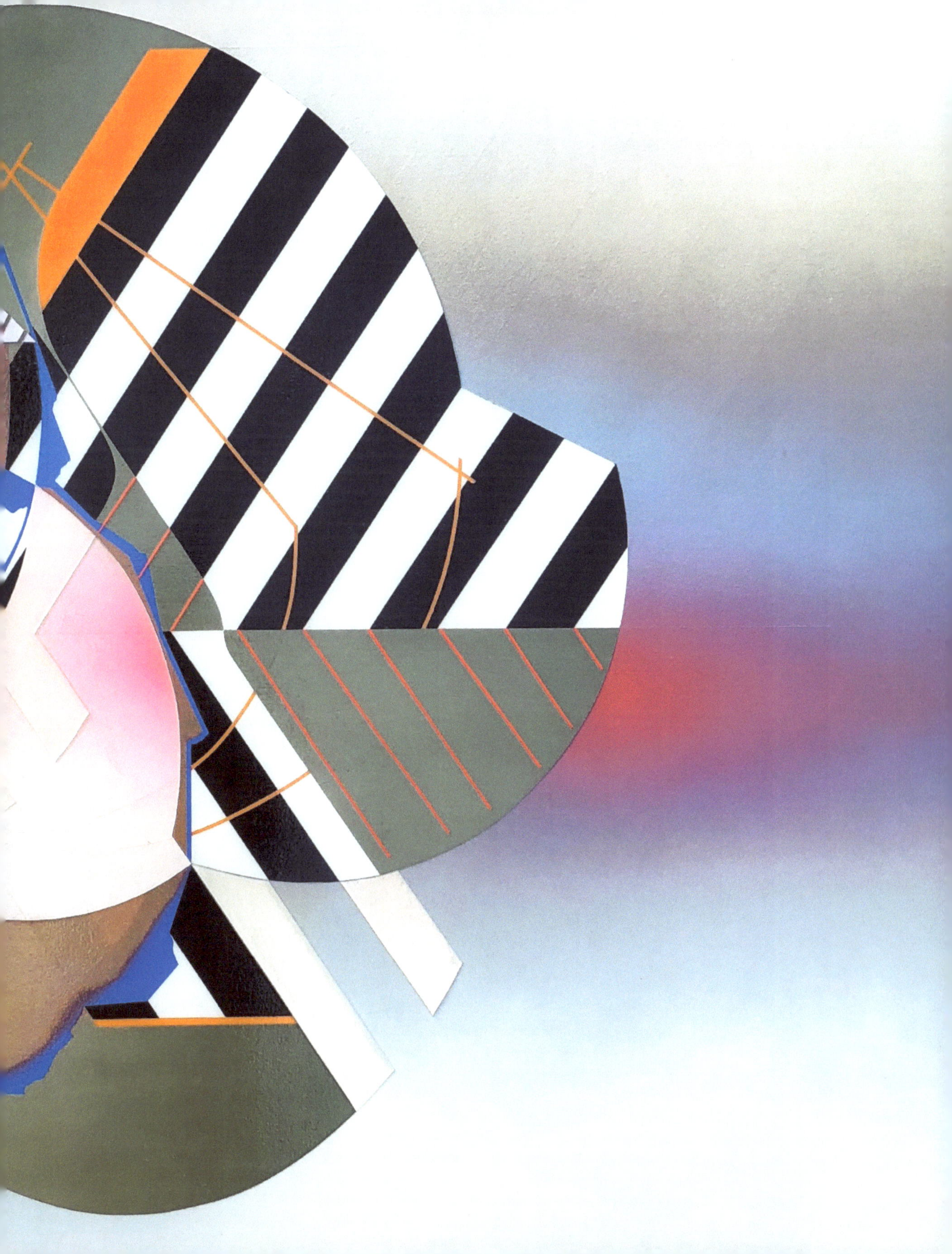

THE DESERT AND SPACE AND NOTHING AND NOTHING

the desert, unlike any other location or terrain, enables a heightened sense of space
the conditions are:
        should I list these yet?
                also, as awareness of space heightens, so does awareness of time

how the desert holds space
mostly in the sum of its absences
unlike a forest, there are no trees
no flora large enough to visually interfere
no structures of an inhabited place
no impediment to the expanse fully defining itself
it's not purely an expanse, though, so it doesn't have the same effect as standing
in the middle of a plain, or floating in a body
of water, the desert's expanse is punctuated by far-off landmarks
faded by atmosphere
            such that the only thing to see is the distance itself
                        distance as object
                        emptiness as stage
                        space as monument
                        math?
there is no movement in the desert
nothing to signify a change in mass or position
no sound
and therefore, no intervals between sounds
            no space between sounds? no rhythm
            rhythm is repeated space between sounds?
      music and memory both use repetition
              but one is stretched out and distorted

every time you access a memory
it changes, but you think it's the original
but does memory feel like music slowed down
to the point where its rhythm is hardly perceptible as music?
what does the span of a life sound like sped up?
there is nothing to measure time

space is a vacuum

in the desert
you are small
you are not only small, but you become aware
that you are small
you are solitary
in a mass of silence / vacuum of sound
the only thing
you hear is
desert plants
you hear your breath
your heartbeat
your internal monologue
being syphoned out
to fill the volume of the desert
here I wanted to talk
about a sense of freedom, a lack of restrictions?
the desert as an emblem of Manifest Destiny?
everyone knows Monument Valley
and property rights
barbed wire
where rockets are tested
the thought comes back distorted
the desert is the passage of time looping into itself
elongated, frozen
appearing all at once
resembling an earth long ago, or far in the future
both ancient and prophetic

there *is* actually sound and movement in the desert
just minimally so, approaching zero
the desert is to movement as John Cage's 4'33" is to silence
in 4'33", silence draws attention to the audience
as I stand here
this is an uncomfortable stage

space is too vast to describe
space is described by numbers here
space (the perception of it) is a concept
          of the distance between masses
          relative to your own body
if I pay close enough attention, I can locate the self
        or the illusion of it? through a progression of observations:
"self" as in, not the ego or a physical self or identity,
          but what appears to be some thinker or observer
                    some center location in the mind, authoring a path
                    some wanderer
among desert plants

observe the breath
observe the sensation of being in a body
the earth pulling it down
                      then thoughts
observe the instant they arise and
thoughts dissolve?
normally you don't pay attention to all of these things but you discover that
          they're all automatic functions
          that feel integrated into a unified "self" of perception/thought
          but while performing this, that unified self splits
                    into (1) the observer and (2) the observed, which I previously thought
                              was me, but was already determined
                    and feels ancient?
          if I could sneak up on automatic functions that I thought were "me"
                    could something else sneak up on myself?
          what is this space, and will it let me

                    crawl out of myself completely

I said there is no movement in the desert
but there is
when I think of the desert, though, I think of land
          what I perceive and experience are
categories
mostly abstract categories
like scale, for example
I can't really describe a specific experience with a desert plant, no
I don't think about a little tiny plant waiting for the wind to come by
when it eventually does, just
                                        scale

because I don't have a lot of thoughts about desert plants
I have thoughts about the geological architecture of the desert
it's the larger-scale things that are important to what I am trying to say about the desert
having a certain set of conditions that make you think more about space
and make you think about time as well as space
what about time
time is the moon or a clock

there's some weird mix of religion and devotion in repetition
if religion sustains itself on tradition and ritual
I get that same feeling with these line drawings
being a slave to exacting the number, the ratio
there is nothing else
except the abstract
immatérial concept of math
                    the mathematical description of space
and the artist devoting herself to that

Agnes Martin could have made these only in the desert
compressing hours of someone's inner life
                              into something that can be
                    viewed all at once, held

we see the human
as the departures from a perfect mathematical line or structure
it seemed to be very old
submitting to something larger than itself
in the desert
you are
repeating the imperceptible, on a grid or in rows
it's the subtleties
that are the subject, the slight deviations
                              what you bring to the stage
the hand's slight deviations from the geometric
                                                            echo
the slight variations in hue
                              between fields of color

in the desert
is math discovered?
          or does it arise simultaneously with our construction of it?

things to do with space:

make it fabric
slump over it
bend it over a bar
pile them
stitched together?

what was initially about time and being
slightly outside of your body
                                        at the end is
            the beginning of
submitting to something larger than you
at the same time you're constructing it
I like this idea of building something
                        like a monument
the sound of math's becoming

an equation to describe
            the elliptical orbit of the moon
            the trajectory of a rocket

which trends toward what? because you *can't* really
hold it, view it all at once
there's always some distant
version of you

because
traced at its center
nothing ever returns

to exactly where it left

- Cooper Johnson
Art Critic and Writer

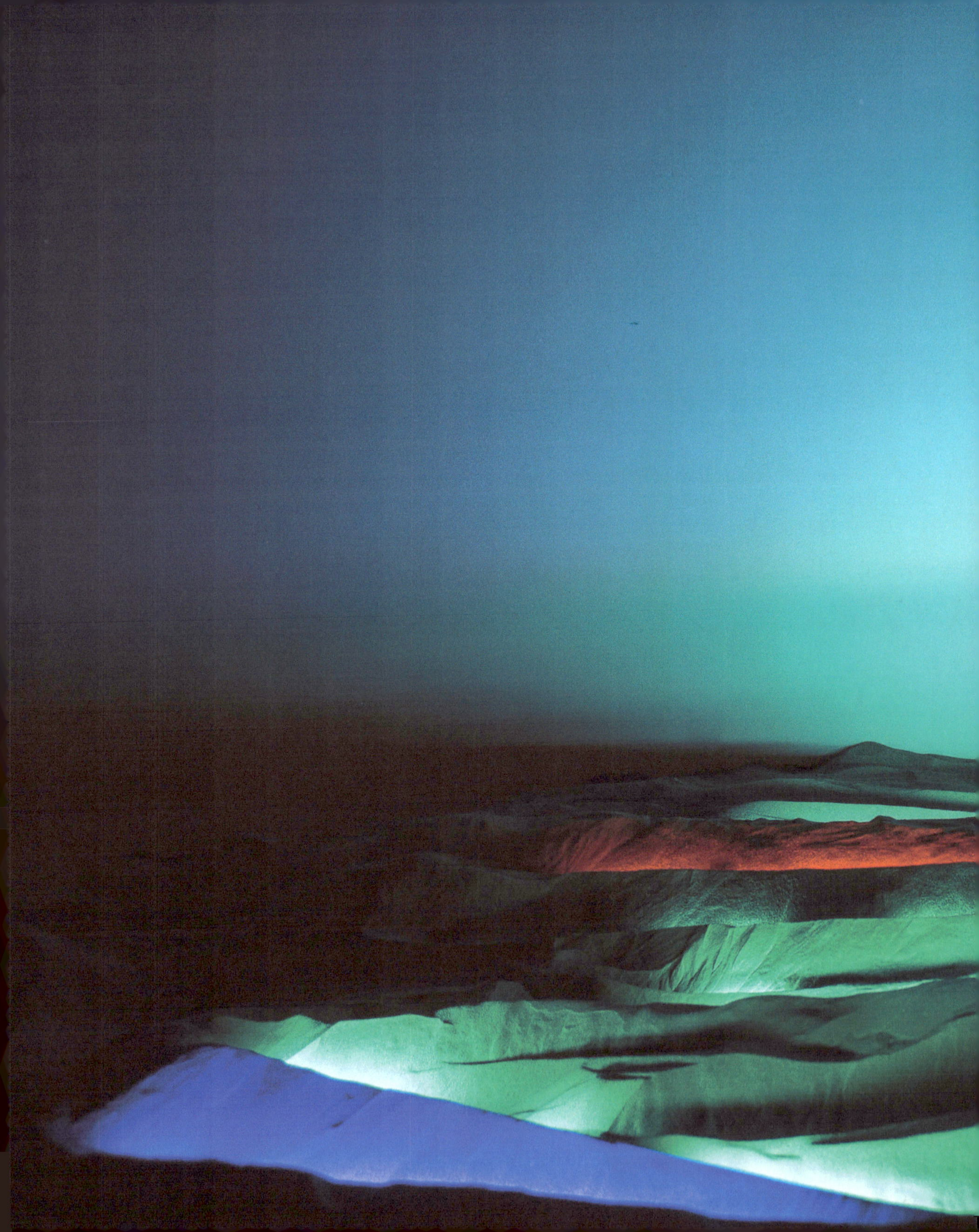

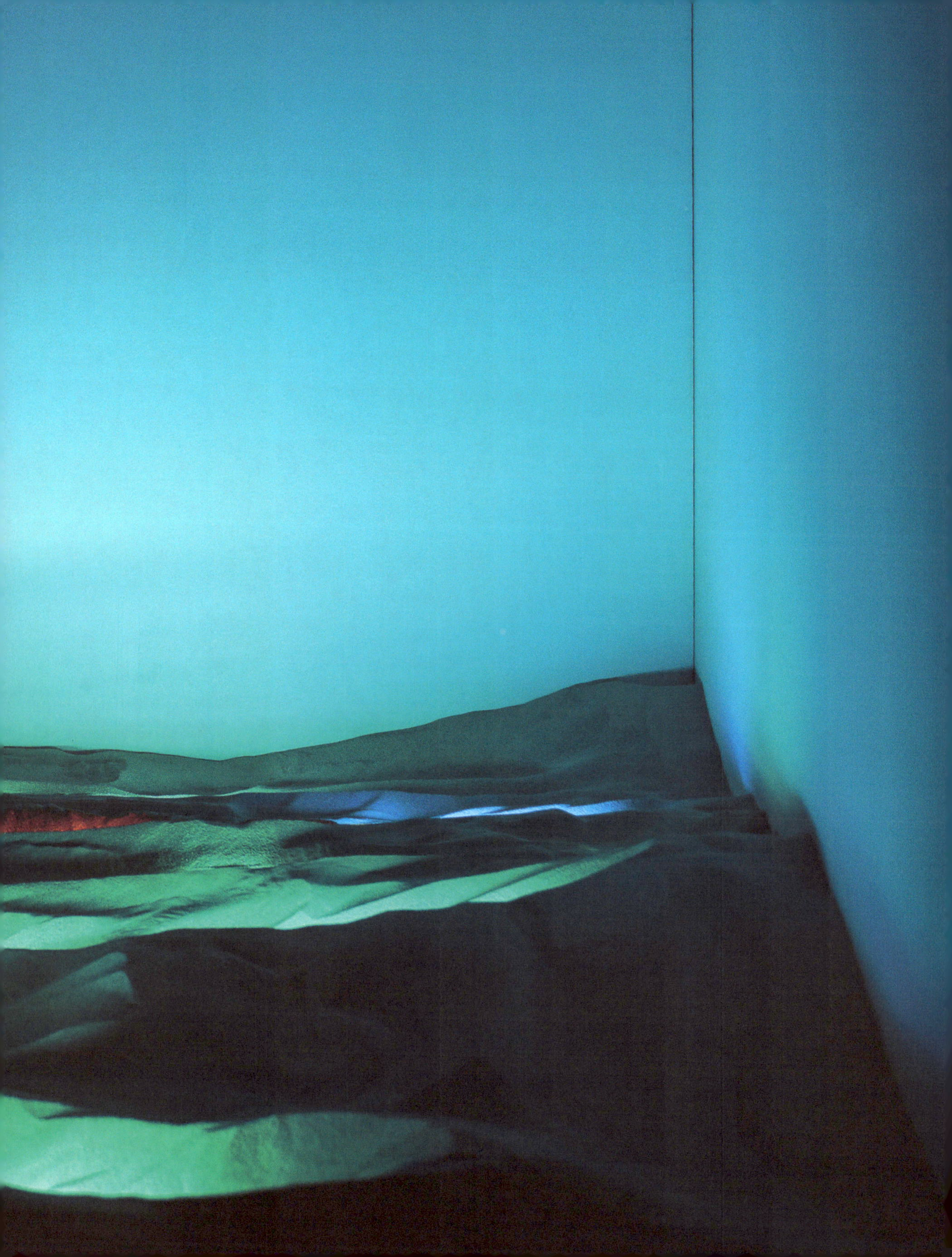

SPACE AND INTELLIGENCE

Space is a universal topic, yet it has many definitions. Its astronomical meaning implies the regions of the universe where there is literally (almost) nothing. On Earth, however, our atmosphere is also a bit like space. Like the vacuum of space, air on Earth is usually transparent to light, does not impede our motion very much, and it doesn't move objects around (unless it is very windy).

It turns out that treating air as space is a very useful approximation. The space-like properties of air have had a huge influence on the evolution of our brains. We evolved receptors in our eyes that can sense radiation—light—at huge distances across space on Earth and into the sky. Our brains contain a structure called the hippocampus in which can be found "place cells" that seem to represent spatial locations that are meaningful to us. Rats (and possibly humans) also seem to have "grid cells" that enable them to measure distance in a regular way. This mastery of understanding space is crucial to any species wishing to survive and dominate the space-like environment on Earth's land mass. One can truly say that we and our fellow creatures are "space beings."

Starting in the year 2000, my colleagues and I started to look at the relationship between intelligence and space in a different way. What would it mean if the space itself was intelligent, in addition to the beings living in it? Could the space learn about its inhabitants? Could it learn to influence their actions? To investigate these questions, we embarked on a unique hybrid project that combined elements of science, art, and entertainment. We created a large-scale interactive exhibit called *Ada:*

Intelligent Space, which was shown to more than 550,000 visitors over a period of six months during the Swiss national exposition of 2002. Our collaborators included scenographers, lighting designers, and artwork and sculptures provided by H. R. Giger of *Alien* fame.

The space was conceived as an immersive environment of computer-controlled light and sound, a sort of *Saturday Night Fever* disco for the twenty-first century. The space featured a hexagonal grid of neon-illuminated floor tiles with pressure sensors, which we designed and built from scratch. It also had an array of cameras and moving lights with a 360-degree projection screen, and microphone arrays that could detect and localize noises. To use the obvious animal metaphors, the floor was Ada's skin, the cameras were her eyes, the microphones were her ears, and the lights were her fingers. Ada's interactions were driven by a cluster of more than thirty computers running a mix of neural network simulations and agent-based software. In a back office resembling a kind of control room, we could monitor and control Ada's operation on a bank of monitors.

Building the space was an arduous two-year project involving a team of about a dozen people. When the six-month operational period started, we juggled the tasks of simultaneously entertaining thousands of visitors, conducting behavioral experiments, and improving the exhibit. The experiments were especially challenging to run, but they yielded interesting results. By altering the ambient conditions and the interactions inside the space, we measured and compared the movements of different groups of visitors. We found that we could subliminally influence their movements and even their emotional responses to the space. By applying reinforcement-learning algorithms, the space could learn to do these things on its own. By the end, we had amassed over 200 gigabytes of data—a quite large data set for its time.

Looking back, the task of conceiving, building, and running the intelligent space was both physically and intellectually exhausting. The practical considerations of ensuring a certain quality of entertainment for paying visitors constantly conflicted with the demands of performing experiments. We were also operating in a new, philosophically ill-defined field. Many hours of argument were spent on the definitions of "inside" and "outside." What did it mean to be "inside" the space, if all organisms interacted with the world "outside"? Bemused initial reactions from colleagues in traditional fields of robotics only added to our collective intellectual insecurity. The experimental results were there, but it was far outside what anyone had done before, and people had difficulty putting it into a mainstream research context. In short, the time was not ripe to be looking at intelligence in the context of environments.

Slowly, perceptions started to change. Initial support came from fields with a more artistic inclination—my first invited talk on the topic was at an architecture conference. A few years later, we were calling Ada an early example of a mixed-reality space. In 2012, ten years after the exhibit closed, I gave a TEDx talk about Ada. It would be another five years before the concept of the Internet of Things entered the collective consciousness, and consumers began turning their living spaces into sensory, intelligent environments. Now, almost another decade later, in public environments everywhere I see echoes of our initial explorations. And I still feel that it is only the beginning.

The work we did on Ada led me to work subsequently on other issues of space—virtual space and how it is processed in the brain. Much later, I started looking at space from a robotic perspective, looking at how robots understand space through vision. But that is another story.

- Dr. Kynan Eng
Cofounder and CEO
iniVation, and Research Group Leader
Institute of Neuroinformatics
University of Zurich and ETH Zurich

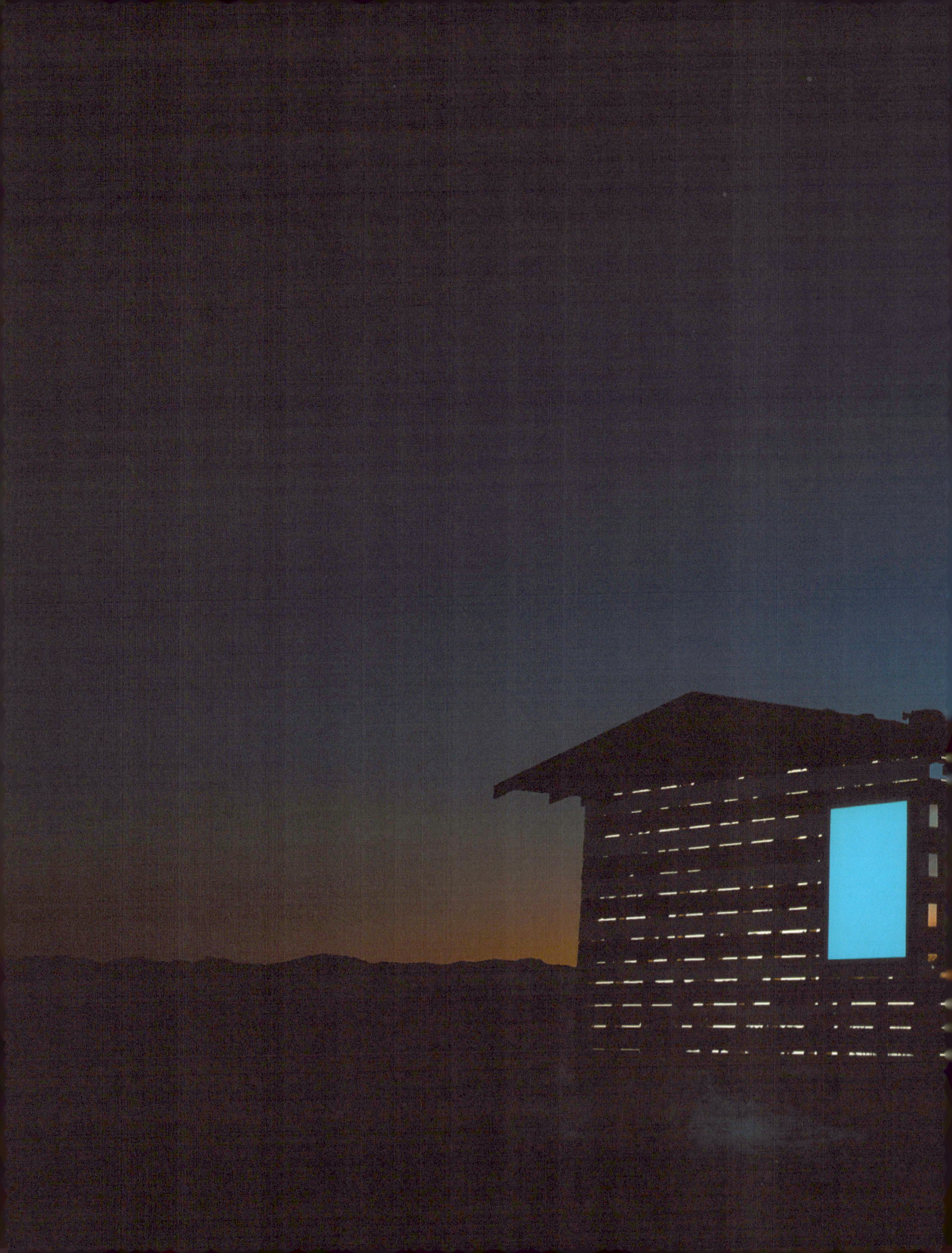

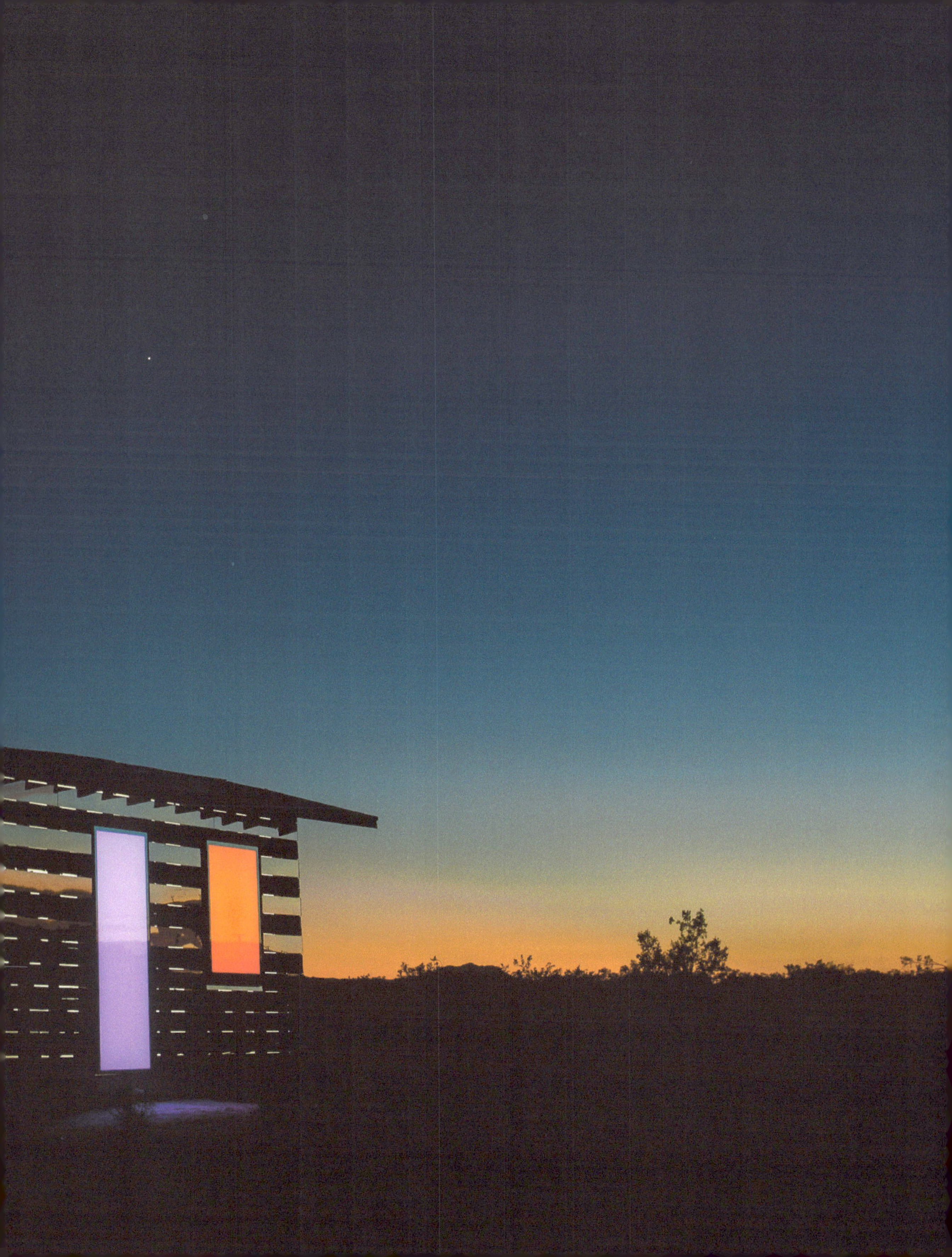

SPACE AND CAPITAL

I've spent my career, more than five decades now, in finance. In early 2000 I founded the investment firm Ridgestone Corporation based in Los Angeles, which invests directly in a wide range of industries via publicly traded securities, real estate, and private equity and venture capital transactions. Like any other profession, finance has its own jargon, not only related to the specific terminologies of capital investment—things like derivatives, swaps, hedging, yields, etc.—but also a more generalized way of speaking, of presentation, essentially the way we communicate when we're communicating about *making* capital investment. It seems to me it was with the advent of the internet revolution, the first wave of dot-com investment, that the word *space* began to be used in a way that, when you stop to think about it, simultaneously connotes a seemingly circumscribed, defined area and yet suggests a kind of boundlessness. We often hear things like, "So-and-so is an expert in the XYZ space," with XYZ being things like biotech, artificial intelligence, streaming, telecom, venture capital, you name it. In that context, you could just as easily substitute words like *field* or *sector*—the "biotech field" or "biotech sector"—yet those words feel more fixed and staid, and in the case of *field*, two- rather than three-dimensional. *Space* on the other hand subtly seems to convey something more dynamic: a nebulous realm of flexible boundaries that sort of sparks the imagination through its conveyance of a sense of almost limitless potential at both the macro and micro levels. I suppose a cynic might say there's nothing more than trendiness to the word being used this way, that it doesn't really mean anything different than words like *field* or *sector*. But I think in its own small, subtle way, *space* in this context seeks to capture something of the spirit of our technological age and its yet-to-be-imagined possibilities.

- Abbott Brown
Chairman and CEO
Ridgestone Corporation

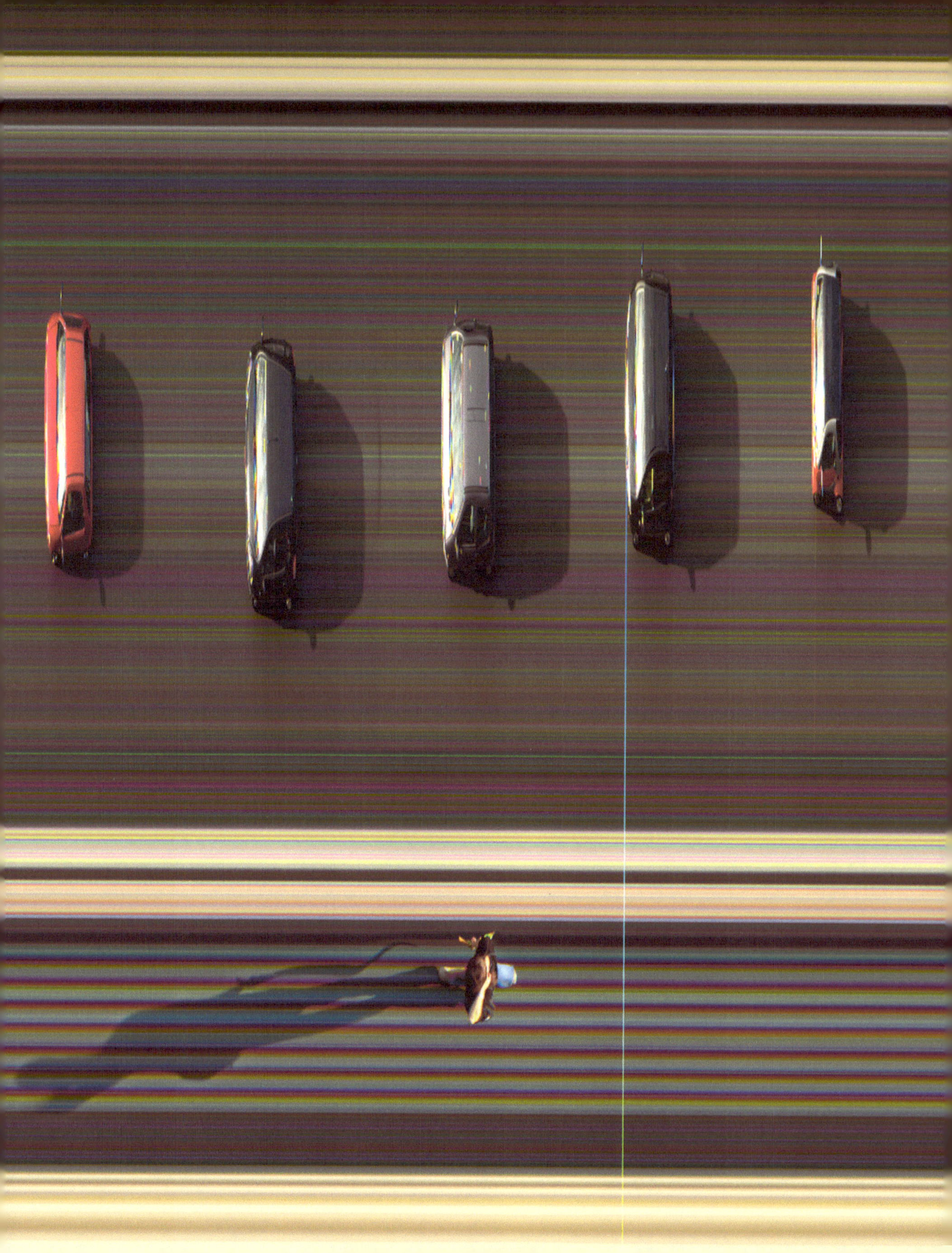

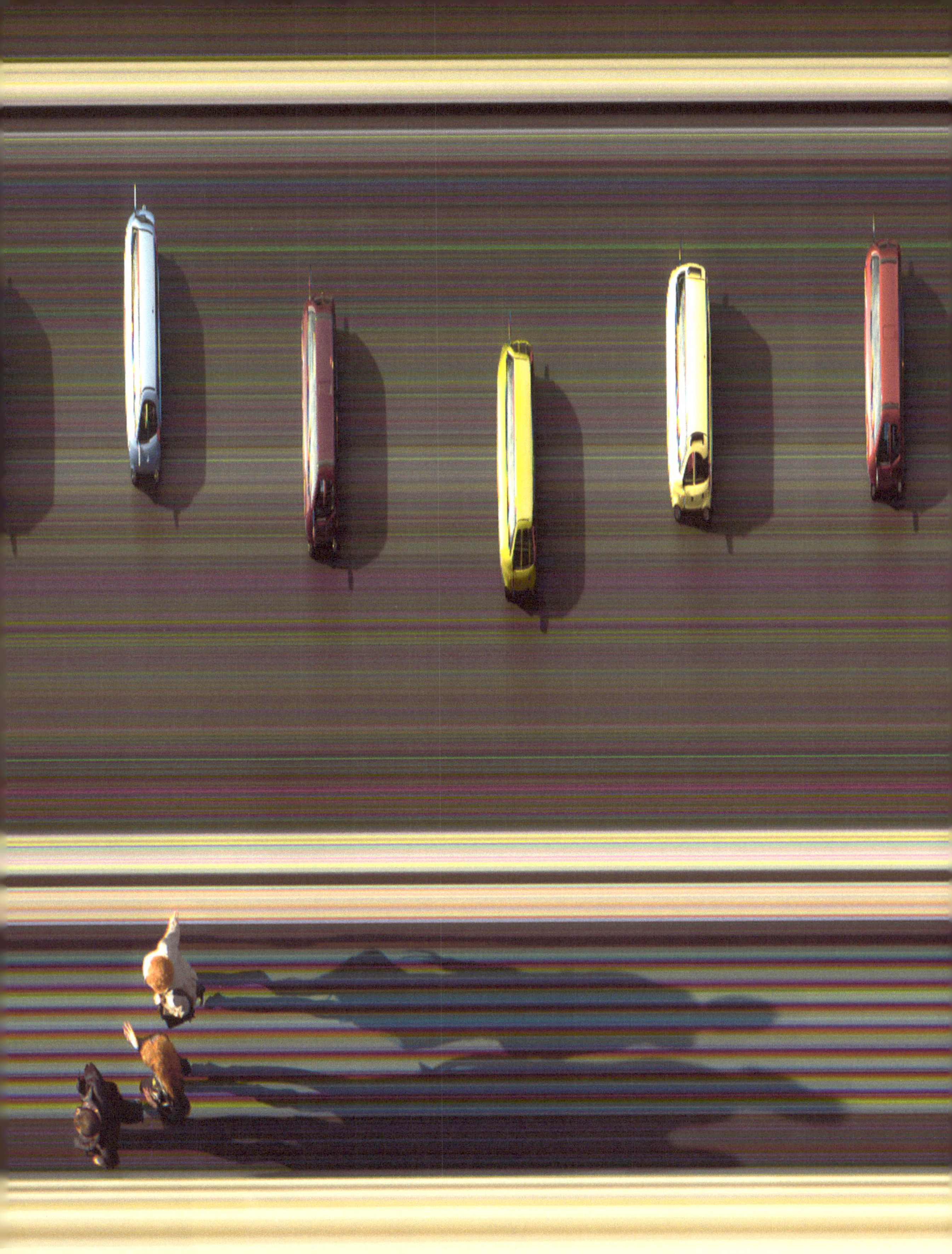

SPACE AND OCEANS

I have seen one. An alien. The first one I saw was at Cocos Island in Costa Rica in June 2018. And then I saw a whole group of aliens near Eua Island in the kingdom of Tonga later that year. One of them, perhaps forty feet in length, tracked me with her fist-sized blue eye as she elegantly glided past, expressing so many things to me in her glance. The encounter was magical and impacted me deeply. I still feel . . . honored.

The Cocos Island alien was a fetchingly fat tiger shark of about eleven feet in length. I knew tiger sharks were often sighted in the area as we had been warned not to dally on the surface near an offshore rock called Isla Manuelita. So I was thrilled to see this gray-striped beauty on my first dive, barely twenty feet below the surface. I steadied my depth, controlled my heart rate, and stared at her in awe as all thoughts in my head ceased to exist. She eyeballed us as she continued onward gracefully, not minding us divers clad in black wetsuits and scuba gear toting large digital cinema cameras.

The Tonga aliens were Southern Hemisphere humpback whales with long, sleek dark gray bodies with white or mottled white and gray pigmentation underneath. You can see little knobs all over the body called tubercles, and if you're close enough you'll notice a single tiny hair protruding from each one. The underside of the humpback shows magnificent long white ventral pleats, and often affixed to various places on the body are hitchhikers in the form of barnacles. The closer you get, the more otherworldly these beauties look. Each whale has its own unique markings, so it's possible to tell them apart if you are able to hold their markings in your brain while absorbing all of the beauty swirling around you at the same time.

These alien sightings occurred while I was working on my documentary *WATSON*, a film about the life of Captain Paul Watson, a man who will stop at nothing to save the oceans and what lies beneath. When I started the film I knew that over 70 percent of our planet is ocean, but I hadn't realized we, as humans, have only explored about five percent of it. Nor did I realize the immediate and alarming extent to which it is in peril. The ocean is wildly overfished, dangerously warm, and polluted with microplastics and other trash that threatens all within it and, ultimately, threatens all of us.

It seems we have explored the topography of the moon and even Mars, yet right here on Earth we've only laid eyes on a small portion of the seafloor. Which is odd considering the ocean impacts our daily lives in a manner that Mars does not. Up to 80 percent of the oxygen in the air we breathe comes from the ocean, making it both an oxygen provider and climate regulator.

More people have been to the International Space Station than to the deepest known part of the ocean, the Mariana Trench. Perhaps it's easier to put someone in space than on the bottom of the ocean where the water pressure is enormous and the darkness is total. Yet it holds many wonderful surprises. Scientists who explore our ocean's depths find new organisms and species, some so unusual they exist beyond the boundaries of current scientific understanding. In fact, more and more "alien" life forms are being discovered the further we care to look, many so wild and weird they rival anything in science fiction.

While filming at Cocos Island, just seventy-five feet beneath the ocean's surface, I was startled by the diversity of what I saw. Purple sea urchins atop unearthly rock formations, orange starfish clinging to colorful coral gardens, metallic jacks in such quantities they appear a mosaic of mirrors swimming past, and parrotfish munching on coral with their full set of hilarious white teeth. Floating freely were rays, pufferfish, snapper, white- and black-tipped reef sharks, hammerheads, and so much more that exceeded my capacity of identification.

I think it is in the larger creatures of the ocean where you can more easily discern alien intelligence. I say "alien" for two reasons. One is that below the surface of the ocean lies a realm that is just as foreign as the moon.

The second is because I cannot fully explain how a humpback whale, for example, can calculate precisely my movement and position in the water, avoiding me within inches, while accounting for our dramatic size difference, not to mention the ocean current and other whales. This type of empathy, awareness, and navigation skill is a mystery in itself.

For days we tracked the same Southern Hemisphere humpback whale mother and calf, occasionally joined by a group of five or six courting males that would come and go, hoping to mate this season or next. The mother allowed our interaction and permitted her calf to play near us. In one instance she and the calf were gliding by so close the mother had to gently lift her massive pectoral fin, or flipper—which is about one-third her body length and the biggest single appendage in the animal kingdom—right over the head of our cinematographer, calculating exactly how close she was to him. If a whale doesn't want you nearby, they are gone in a flash. It's hard to imagine how an animal that large can disappear so quickly, leaving you starring into a blue abyss wondering what happened and whether you imagined a forty-foot-long creature only seconds before. When a whale does permit interaction, it's like nothing else I have experienced here on planet Earth. Often when we talk about space it means something vast or distant, yet these encounters showed me there is much to be gained by decreasing space . . . by simply moving closer to things to experience their beauty and character.

The Southern Hemisphere humpbacks migrate more than five thousand kilometers from Antarctica to arrive midwinter in Tonga and other places near the equator to give birth, nurse, and mate, returning south in the spring. They also play and sleep. Sleeping whales are fascinating to see. They shut off one hemisphere of their brain and can drift along vertically, looking like elegant wise sentinels . . . another truly gorgeous expression to behold.

Speaking of expressions, it is said you should not anthropomorphize animals and creatures of the sea by projecting human wishes and emotions onto them. It was once a fundamental law of animal science. Yet a closer look at whale behavior shows this to be untrue. Whales are socially intelligent, likely more so than us, and they have better built-in navigation. They are mammals that nurse their young and can raise them in groups, live in small whale societies, hunt in packs if they wish, have their own language communicated through clicks and song, and they are playful and quick learners. They show love, affection, and empathy. They also grieve. And some of them may be trying to send us messages, as some believe an orca tried to do in 2018 when she pushed her dead calf around on the surface of the ocean for seventeen days in the waters of the Pacific Northwest.

Captain Paul Watson says we live on planet Ocean and not, actually, on planet Earth. After all, think of all those beautiful shots of Earth from space . . . what do you see the most of when you look at the big blue marble? Oceans. By looking more closely at our resident aliens, we have a chance to visit "oceanic space" again and again right here on Earth.

- Lesley Chilcott
Filmmaker

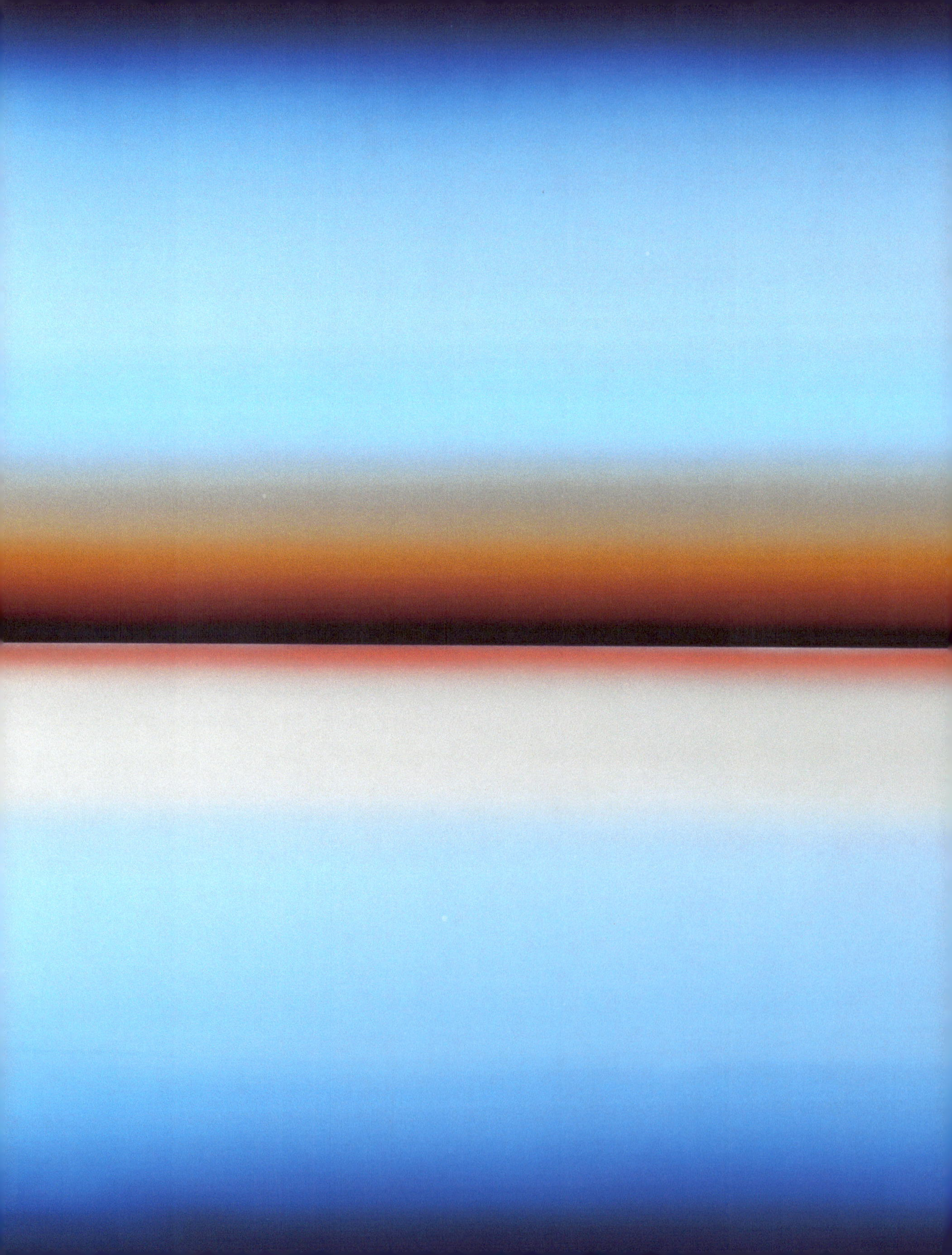

SPACE AND EXHIBITION

I am an object that time and energy gathered in space. The laws of physics govern my spirit and gather in an invisible block my body of flesh.

– Clarice Lispector[1]

In my field of work as a museum professional and curator, I grapple with the exhibition space, navigating its complicated and exclusionary history while seeking new strategies to rethink its traditional spatial constraints. The "white cube" has functioned as a site of cultural production, canonization, and validation, with its hegemonic power structure largely remaining male and white.[2] The historical inability of museums and cultural institutions to value and celebrate the artistic contributions of women and artists of color has led to the prevalence of exhibition spaces that are elitist, exclusionary, homogeneous, and unrepresentative of the current social and political landscape. Yet the exhibition space can also be a site of the creation of new social and political realities, and its dormant potential needs to be reclaimed. I think of Shana Mabari's artistic practice, which grapples with the ways we process and perceive our surroundings. Using art to explore scientific principles, her unique multidisciplinary practice is rooted in playful exploration of perception and space. Inspired by her practice, which continuously reflects on how we perceive our spatial reality, this essay aims to reflect on new models for exhibition making.

The inequality of past exhibition modes has recently been challenged, with new proposals emerging that seek to create more equitable, diverse, and inclusive exhibition spaces. Current demands for more inclusive exhibition spaces has led some museums to take a critical look at their collections and the holes in their collecting policies. One example is the Baltimore Museum of Art's plan to deaccession works by white male artists and to use the funds to accession works by women and artists of color.[3] This art-historical reexamination corrects the historical exclusion by reinserting underrepresented artists into the museum collection and broadening the expanse of the white cube. Curators will thus be able to draw upon more representative collections and to use these works to tell a different art-historical narrative. Other museums are following suit in a critical analysis of their collecting policies and a realignment of their priorities.

It's important to acknowledge the detrimental effects of exclusionary exhibition spaces when sites of cultural production are only catering to a homogenous audience. Griselda Pollock reflects on how systemic erasure has a dual effect: erasing the past, which necessarily results in a missing future.[4] At this moment we are dealing with the effects of a missing future that remains unfulfilled by the historical failure of our cultural institutions to create exhibition spaces that reflect the communities they serve. The erasure caused by this systemic exclusion of Latinxs, for example, has played a part in the current dehumanization of brownness and *latinidad,* in which we have been vilified, persecuted, attacked, and caged. A 1994 report of the Smithsonian Institution Task Force on Latino Issues concluded that "the Smithsonian Institution, the largest museum complex in the world, displays a pattern of willful neglect toward the estimated 25 million Latinos in the United States . . . the Institution almost entirely excludes and ignores Latinos in nearly every aspect of its operations."[5] This historical exclusion has left a lasting impact and has created the need for a new generation of curators and museum professionals to radically transform many of our cultural institutions.

How does this futurist exhibition space exist and function? One transformative example is the Getty's far-reaching Pacific Standard Time LA/LA (2017–18), which provided grants to more than seventy institutions in Southern California to present exhibitions exploring the cultural connections of the Los Angeles area and Latin America. The ambitious program created a platform for the radical transformation of exhibition spaces that created new scholarship and recovered marginalized voices. Its flaws notwithstanding, it also helped to reveal the many cultural institutions in Southern California that have not reflected Latinx culture in their collections, exhibitions, programs, or among the ranks of museum staff.

A recent study revealed that PST contributed roughly $430 million to the region's economy.[6] The substantial economic impact derived from an exercise in transforming the exhibition space leaves one to wonder what other futures have we been robbed of at the hands of the exclusionary exhibition space?

A moment is needed to reflect on the exhibition space as a site of dormant and unrealized potential that, once fully activated, can unleash transformative energy. Originating as cabinets of curiosity, the exhibition space has largely been allocated to the white heterosexual male gaze. Such myopia has led to drastic art-historical voids. The need to (re)imagine a new baseline for exhibition-making is pressing. Barriers that have prevented these spaces from welcoming women, people of color, and queer communities need to be erased, which necessitate the dismantling of white heterosexual male supremacy. Helen Molesworth suggests the possibility of a reorganized institutional dynamic that can, in turn, realign societal dynamics and power structures as well.[7] Exhibition spaces serve as important contact zones at the juncture of ideas and cultural creation.[8] Spectators can encounter new ideas. These spaces are also the nexus of cultural creation that at times lead to new ways of seeing and understanding. I think of Mabari's *Afterimage* (2011), in which she toyed with images that continued to appear after the original exposure stimulus.

Historically, exhibition spaces in museums, cultural institutions, and galleries have been white, male, and hetero-centric, but the future can be radically different. During this critical social and political juncture, it is necessary to challenge the entrenched power dynamics and to offer up space to those who have been excluded so as to further empower their participation in the process of cultural production. By rethinking the exhibition space, we invite the reconceptualization of one crucial form of cultural production. When the marginalized voices that have been excluded from this space are welcomed into the process of cultural production, the shift will be transformative. The chance is more urgent than ever.

- J. Raúl Guzmán
Curator
Academy Museum of Motion Pictures

1 Clarice Lispector, *A Breath of Life*, trans. Johnny Lorenz (New York: New Directions, 2012).

2 For a historical overview of the development of exhibition spaces, see Tony Bennett, "Exhibi-

tion, Truth, Power: Reconsidering 'The Exhibitionary Complex,'" in Quinn Latimer and Adam Szymczyk, eds., *The Documenta 14 Reader* (Munich: Prestel, 2017).

3 See Julia Halperin, "'It Is an Unusual and Radical Act': Why the Baltimore Museum Is Selling Blue-Chip Art to Buy Work by Underrepresented Artists," Artnet News, April 30, 2018: https://news.artnet.com/market/baltimore-museum-deaccession-1274996.

4 Griselda Pollock, "The Missing Future: MoMA and the Modern Women," in Cornelia Butler and Alexandra Schwartz, eds., *Modern Women: Women Artists at the Museum of Modern Art* (New York: Museum of Modern Art, 2010).

5 Smithsonian Institution Task Force on Latino Issues, "Willful Neglect: The Smithsonian Institution and U.S. Latinos," May 1994. Accessed online at https://siarchives.si.edu/sites/default/files/forum-pdfs/Willful_Neglect_The_Smithsonian_Institution%20and_US_Latinos.pdf.

6 Henri Neuendorf, "Pacific Standard Time Added $430 Million to Southern California's Economy, Report Says," Artnet News, May 1, 2018: https://news.artnet.com/art-world/pacific-standard-time-economy-1276865.

7 Helen Molesworth, "How to Install Art like a Feminist," in Butler and Schwartz, eds., *Modern Women.*

8 For "contact zones," see Mary Louise Pratt, "Arts of the Contact Zone," *Profession* (1991): 33–40.

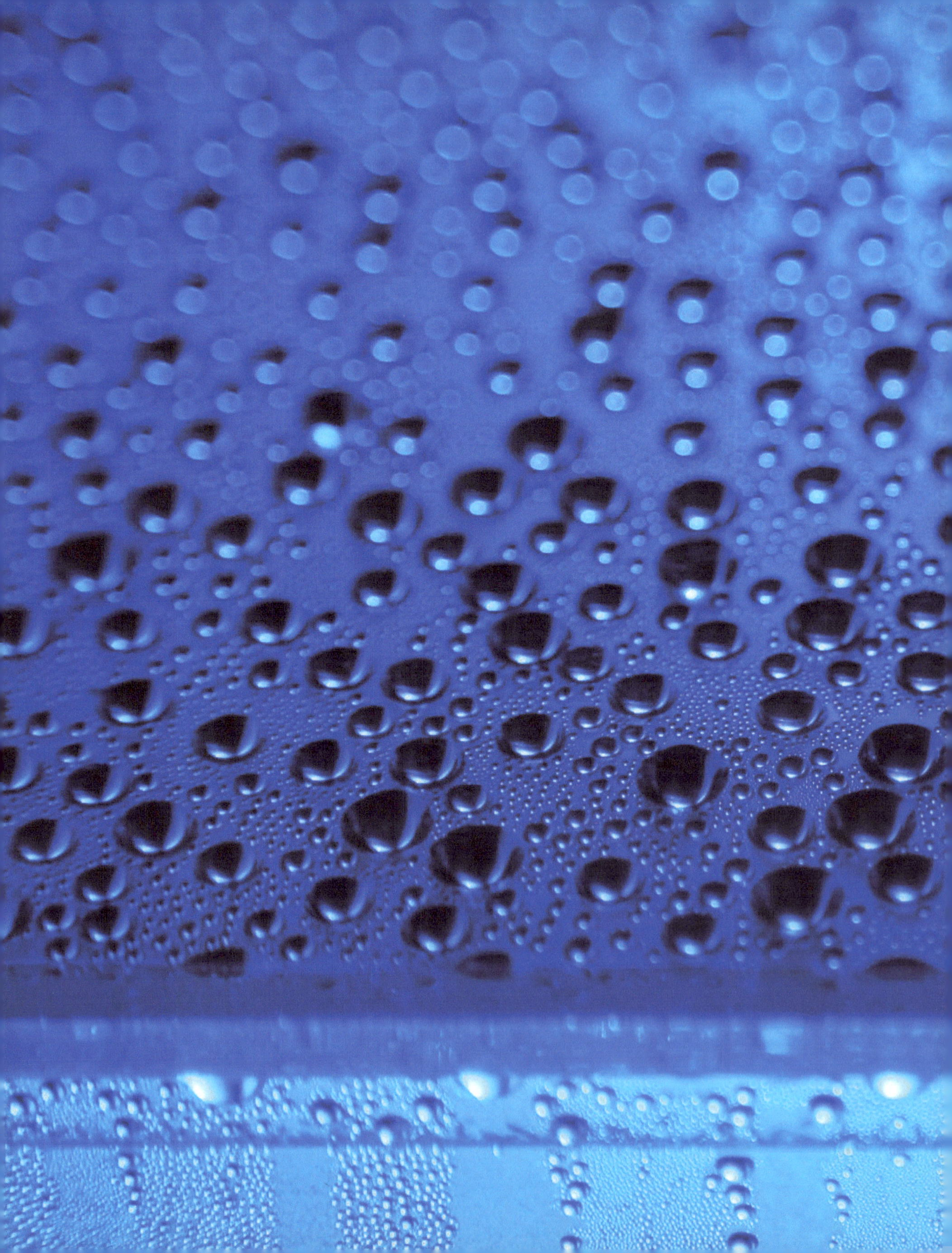

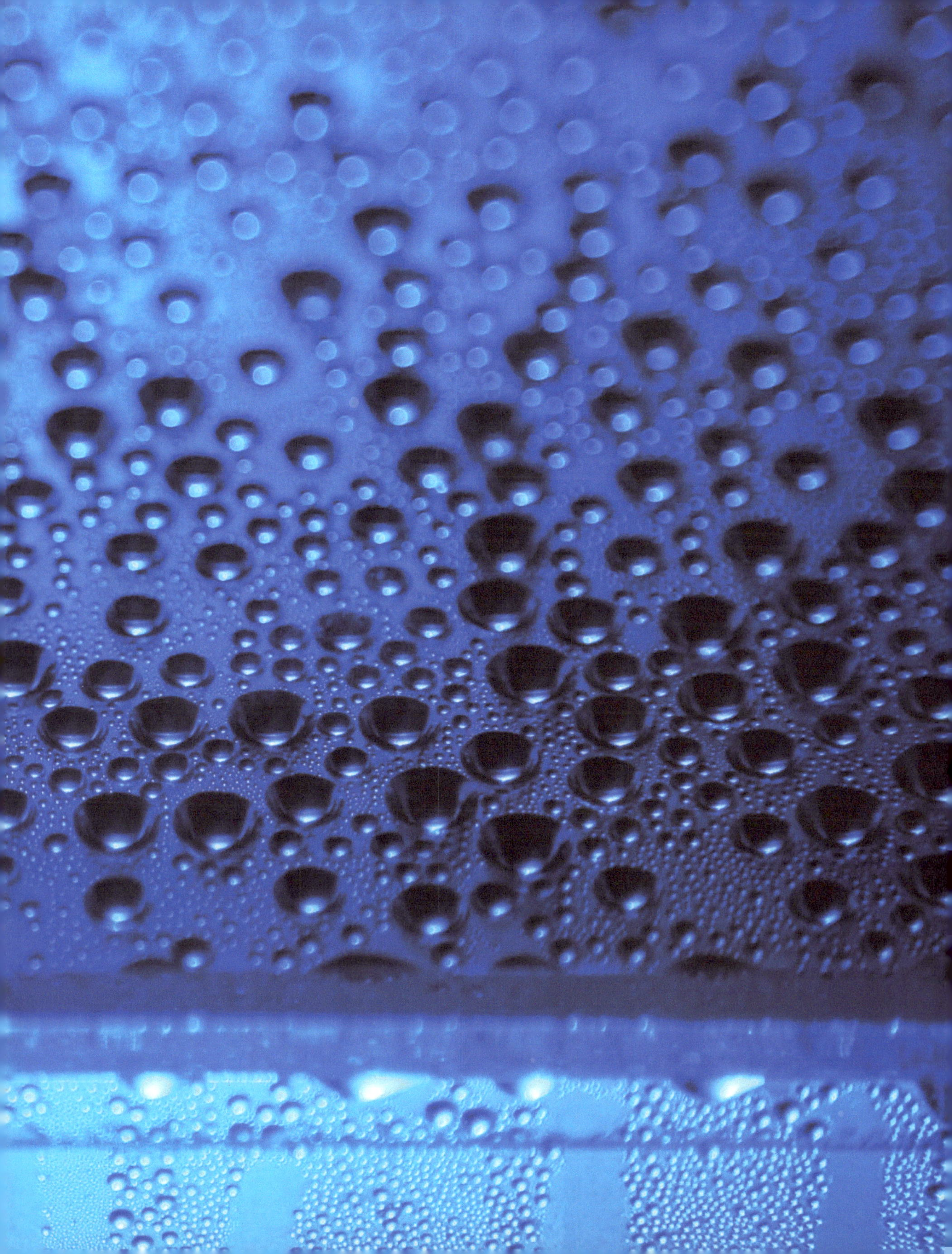

[EL] ESPACIO Y SU LÍMITE

No había reparado en la frecuencia con la que utilizamos la palabra—o el término— *espacio* en nuestro vocabulario hasta que Shana Mabari lo puso en relevancia. A partir de ese momento cada vez que me refiero a los *espacios* expositivos donde trabajo, me resuena la facilidad con la que hacemos uso del término, quizá incluso hasta sienta algo de incomodidad por no estar refiriendo el uso correcto de la palabra. Pero desde luego ya está muy incorporada en nuestro lenguaje y podría atreverme a decir que, sobre todo, en el mundo del arte contemporáneo.

El *espacio* se utiliza para denominar: un lugar donde se realizan exhibiciones, la distancia entre las obras expuestas, la utilización del lugar por parte de los artistas y curadores, como metáfora explicativa de los procesos creativos, en títulos de las obras. Trayendo a colación la escultura—por el hecho de que es el medio principal de Mabari—también me puse a reflexionar sobre el concepto del *límite*, el límite *en* el espacio y *del* espacio; físicamente podemos entender un espacio haciendo uso de sus límites, conceptualmente la relación se convierte más difusa y es en ese espacio nuboso donde intuyo que la escultura subsiste con el resto de objetos que pueden habitar un *espacio*.

¿Qué diferencia a una silla y una mesa colocados en el mismo lugar donde una escultura? ¿Cuál es la diferencia objetual entre ellos, es inherente o también depende del uso del espacio que como objetos terminan reclamando? Desde mi perspectiva, es el *límite* el que juega un papel determinante para discernir la diferencia entre un objeto y una escultura, entendiendo la escultura como un objeto de arte. La funcionalidad en este caso está relacionada con el límite de las cosas en su espacio, una mesa y una silla tienen un límite de uso específico, mientras que una escultura no conoce ese límite, por que una mesa y una silla apiladas una sobre otra, podrían convertirse en una escultura. La escultura utiliza su espacio dentro del espacio donde se coloca y bifurca (o pervierte) los límites de ambos, del contenedor y de su contenido—es la escultura como objeto de arte donde podemos encontrar la simbiosis entre el espacio y su límite, entre el límite del espacio y su correlación con los demás objetos que coexisten en un mismo derredor. Pero aquí nos topamos con una complicación que quizá no tiene solución, salvo empírica científica: ¿qué dota de ese carácter de indefinición limitativa a un objeto, el mismo objeto en sí o la carga simbólica que empleamos para justificar su existencia?

El espacio se convierte en un término más peliagudo cuando nos salimos de la caja blanca, del confortable y seguro *espacio* de una sala de exhibiciones y pasamos a lugares más extensos, como sucede con las esculturas públicas. Sin embargo, es común oír referirse a los lugares creados alrededor—o algunas veces dentro—de una escultura pública como el *espacio* escultórico. ¿Dónde está el límite de una escultura como los monolitos en el desierto de Richard Serra? ¿Cómo se relaciona el espacio con las *Siete Montañas Mágicas* de Ugo Rondinone en Nevada? ¿Deja de existir el espacio solamente por que no las contiene un contenedor arquitectónico? No lo creo. El espacio y el límite vuelven a jugar esta relación bilateral e intercambiable pero dentro de un concepto más amplio, que no solamente tiene que ver con el espacio que ocupan, sino con la conciencia que nos generan al enfrentarnos a ellos como objetos de arte. Quizá el placer que sentimos al encontrarnos con estas esculturas—al formar parte de su *espacio*—tenga que ver precisamente con la invitación ontológica que nos ofrecen al dejarnos ser parte de un espacio más grande, en el que somos más pequeños y podemos fundirnos en esa relación bilateral entre nosotros, el espacio y sus límites. Es decir, dejamos de tener un espacio por que dejamos de tener límites. ¿Es esa la función ulterior del objeto escultórico? Es probable. Por que científicamente el espacio tiene un límite, pero en el común de nuestro universo, el espacio es algo que juega a ser infinito, entonces una escultura, bajo estos supuestos, forma parte de un infinito que conceptualmente aprehendemos, aunque físicamente apreciemos y hasta toquemos su límite.

- Javier Aparicio
Director, Art Projects Ibiza

[THE] SPACE AND ITS LIMITS

I had not noticed how often we use the word—or term—*space* in our vocabulary until Shana Mabari put it to me. From that moment on, every time I refer to the exhibition *spaces* where I work, the ease with which we make use of the term resonates; maybe I even feel some discomfort as I am probably not referring to the correct use of the word. But of course, it is already deeply incorporated in our language, above all I would dare say in the contemporary art world.

[The] Space is used to name: a place where exhibitions are held, the distance between the exhibits, the use of the place by artists and curators, in titles of the works, and as an explanatory metaphor for creative processes. Thinking about sculpture—due to the fact that it is Mabari's principle medium—I also began to reflect on the concept of the term *limit*, the limit *of* space and *within* space. Physically we can understand a space using its limits; conceptually the relationship becomes more diffuse, and it is in that cloudy area where I intuit that sculpture coexists with all the other objects that can inhabit a *space*.

What differentiates a chair and a table placed in the same place as a sculpture? Is it inherent or does it also depend on the use of space that objects end up claiming? From my perspective, it is the *limit* that plays a decisive role in discerning the difference between an object and a sculpture, understanding sculpture as an object of art. Functionality in this case is related to the limit of things in their space; a table and a chair have a specific usage limit, while a sculpture does not know that limit, because a table and a chair stacked one on top of the other could become a sculpture. The sculpture uses its space within the space where it is placed and diverts (or perverts) the limits of both the container and its content—it is the sculpture as an object of art where we can find a symbiosis between the space and its limit, between the limit of space and its correlation with other objects that exist in the same area. But here we come across a complication that perhaps has no solution, except for the empirical. What gives this character of limited undefinition to an object: the same object itself or the symbolic load we use to justify its existence?

Space becomes a trickier term when we step outside the white cube, the comfortable and safe space of an exhibition hall and move to more expansive places, as happens with public sculptures. However, it is common to hear reference to the places created around—or sometimes within—a public sculpture as the sculptural *space*. Where is the limit of a sculpture such as in Richard Serra's desert monoliths? How does space relate to *Seven Magic Mountains* by Ugo Rondinone in Nevada? Does the space cease to exist just because it is not enclosed inside an architectural container? I do not think so. Space and limit replay this bilateral and interchangeable relationship but within a broader concept, which not only has to do with the space they occupy but with the awareness they generate when facing them as objects of art. Perhaps the pleasure we feel when encountering these sculptures—being part of their *space*—has to do precisely with the ontological invitation they offer by allowing us to be part of a larger space, in which we are smaller and can merge into that bilateral relationship between us and the space and its limits. That is, we stop having a space because we stop having limits. Is that the subsequent function of the sculptural object? It is probable. Because scientifically space has a limit, but in the universal common usage of the concept, space is something that plays at being infinite, so a sculpture under these assumptions is part of an infinity that we conceptually apprehend, even if we physically can appreciate and even touch its limit.

- Javier Aparicio
Director, Art Projects Ibiza

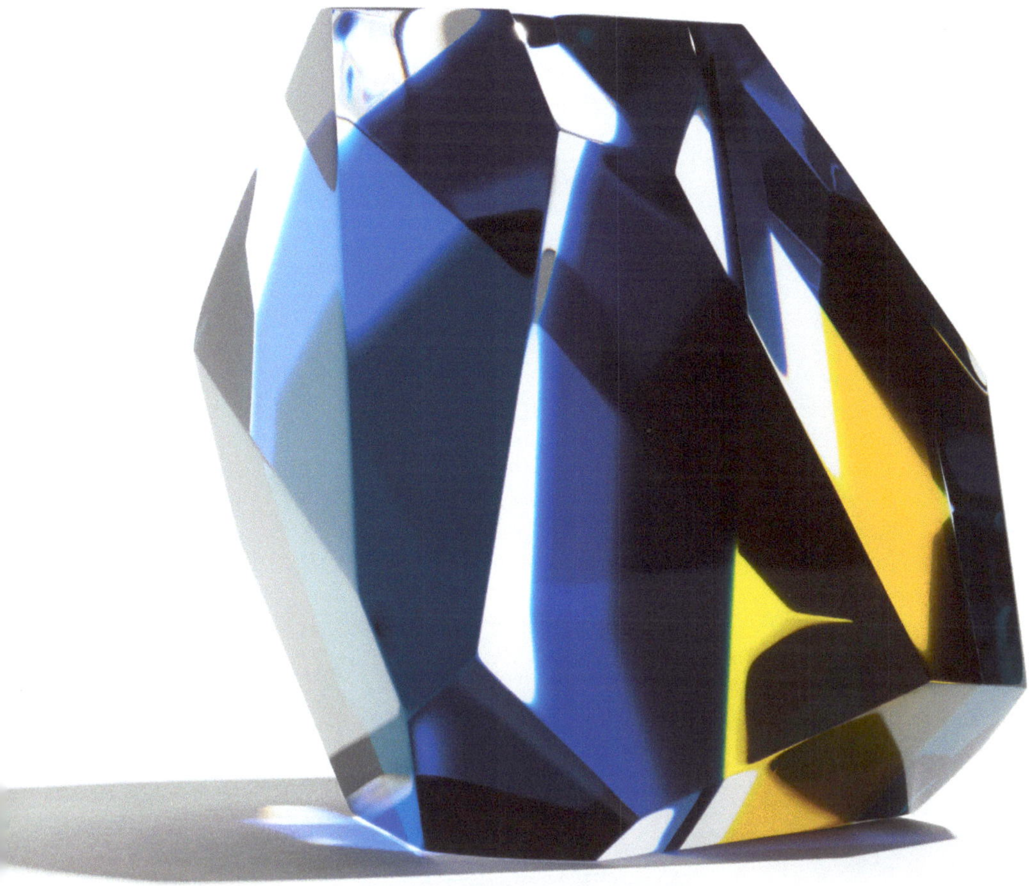

SPACE AND BLUE

The hummingbird-like vibration of blue burst forth in Fauvism, where it was released from its prison in the sky and then could be a tree, a face, grass, or anything. In 1901 Picasso chose blue not just for one painting but for a whole period of his work. Never before had an artist executed an entire monochromatic series of canvases using tonal variations of a single color.

In 1917 Einstein speculated upon a new form of light hitherto unseen. White light, such as light from the sun or any other conventional source, contains the various wavelengths of all the colors of the spectrum, but none can be appreciated in isolation because together they are incoherent. Like a crowd of people, some colors run, some walk, and a few saunter, and all are out of phase with one another. Einstein theorized that under certain conditions light could be emitted from excited atoms in such a way that the waves would fall into lockstep and travel through space much like a drill-perfect, goose-stepping army on parade. If light could be tamed and forced to march in such military precision, then this light, according to Einstein, could be of only one pure color, which would be determined by its source, the emitting atom. Forty-three years later in 1960, Theo Maiman generated the first laser light. Laser, which is the acronym for Light Amplification by Stimulated Emission of Radiation, is the actualization of Einstein's 1917 speculation.

The striking feature of laser light is that it is coherent; it doesn't diverge, and it is pure. Blue laser light cannot be anything but blue. Passing it through prisms, filters, or different media cannot change its color fastness. Picasso's eccentric adherence to a monochromatic scheme sixteen years before Einstein spoke of it, and almost sixty years before Maiman made it real, foreshadowed this immutable new kind of light, and Picasso chose blue—the color of high energy—rather than green, yellow, or red.

Art supply stores at the beginning of the twentieth century had to increase their orders of cobalt and cerulean pigments as one artist after another attempted to outblue the other. In a declaration that couldn't have been made in earlier art periods, Franz Marc, an early abstract painter, declared: "Blue is the masculine principle, robust, and spiritual." The German Expressionists decided collectively upon the name Blue Rider and made a large number of paintings whose dominant color was blue. Paul Klee was a member of that group, and a little over a decade later he and three others formed a successor group to the Blue Rider called the Blue Four. In 1910 a group of Moscow artists called themselves the Blue Rose. Frenetic blue began to appear with a vengeance as if making up for lost time. Late in the 1960s, Yves Klein, like Picasso before him, created a series of all-blue paintings. These flat canvases were covered exclusively with several layers of pigment he called International Klein Blue. Later in his career, he progressed to painting with "living brushes": nude models who covered themselves with fresh blue paint and pressed and wiggled against blank canvases under Klein's direction. This artist reaffirmed the primitive sexual nature of blue without numbers or words.

But for all the blue expended in [twentieth-century] art, the "Blue Award" must go to Matisse, the Fauvist, who best captured the essence of blue. In his famous 1909 work *The Dance*, a vibrant, monochromatic lapis lazuli background provides the atmosphere for the wild, circular dance of five Dionysian maenads. While previous artists such as Degas, Renoir, and Toulouse-Lautrec represented the vitality of the dance, few artists had ever painted a circular one.

A few years after Matisse's painting, nuclear physicists discovered that life itself is based on the carbon atom, which has in its outermost orbital four negatively charged electrons whirling about the stationary, positively charged nucleus. These four dancing electrons are always looking for four more to join them. When their entreaties are answered, they begin to build the interlocking chain-link pattern of life. In Matisse's version, there are five dancers, but there is a break in the hands of two of them as if inviting a sixth or more to participate. The immense deep-blue background is quite startlingly prescient because physicists have discovered since the painting was made that the representative color of nuclear energy is blue. The beautiful and awesome photos taken of radioactive piles at nuclear energy plants are a familiar image of their power. Rather than the roaring red glare from the interior of Bessemer blast furnaces of the nineteenth century, the ultimate image of energy in those times, the silent mysterious blue Cerenkov emanations of a nuclear pile in [the twentieth century] supersedes all previous images. The dance of carbon's electrons and the dance of Matisse's maenads is one and the same dance; it is the dance of life.

Perhaps assigning gender characteristics to atomic particles such as negative electrons and positive protons is not so farfetched. After all, the act of copulation itself, with its implications of creation, is often referred to as blue, and movies containing explicit sex are called blue movies. Music, too, recognized the power of blue when, with unerring perspicacity, a new movement called jazz, wilder and more energetic than its classical cousin, arose from the birth of the blues. Ellsworth Kelly, the field painter of the 1960s, offered a belated recognition of the reversal of the spectrum. In his 1966 rendering of the familiar rainbow, entitled *Blue, Green, Yellow, Orange, Red*, he placed blue to the right and red to the left. This subtle artistic change is the result of a dramatic turnaround in our perception of the color blue.

One year after Matisse painted his incredible blue sky for *The Dance*, Einstein finally solved the problem troubling scientists for centuries; that of the exceedingly basic question: Why is the sky blue? It had gone unanswered by generations of physicists, but not for lack of trying. In 1910 Einstein, building upon Lord John Rayleigh's work, published a paper concerning "critical opalescence" that explained in detail and complex equations the physical basis for the phenomenon of the sky's blue color.

Matisse, the sophisticated colorist, luxuriated in the sumptuousness of variegated light. Einstein, the enchanted child-man, changed the shape of our minds forever because of his incessant curiosity about light and color. Both in their own way assisted light to claim the crown as the rightful heir to the throne of reality.

- Leonard Shlain
Author
Art & Physics

SPACE AND ART

The space of silence, the space of meditation. The space between things: between thoughts and ideas, between objects, between projects. The space with minimal distractions or obligations. The space with zero rules or concrete definitions. The space of creativity. This is the space of art thinking, art doing, art making. The space of art.

This space is as essential to the creative person as air and water. It is the portal if you will (to borrow from science fiction) that connects the artist to the universe of everything that is beyond the individual self, and it is through that connection that the artist becomes a channel, a conduit for ideas and understandings to pass through. What is revealed in this space is unique to the artist and to the situation, to that particular time and space.

The meridians of creativity passing through the artist's body trace trajectories that lead to undiscovered places, terrains unknown to others or even to the artist herself.

The only way to imagine it is to imagine the unimaginable, like imagining the space of the cosmos. It is the space between love; the space between sadness and reflection. The space between life and death.

It is an intentional space, though it is not by definition a physical space; it may be a mental space of silence in sound, in thought, in action.

But the physical embodiment of this space for the artist is most often the artist's studio, the sanctuary where she can enter that vulnerable space of (creative) receptivity.

It is space as absence: the absence of inhibitors, of structure, of laws and rules and expectations. The absence of obligation or goals or ambition. The absence of fear or intimidation.

But it is also space as abundance: the abundance of unbound love, of freedom, of joyful anarchy. The abundance of time and possibility. The abundance of courage.

It is here where the space opens up between ideas, and it is in that space where opposites find connection and logic seems divided (or logic portrays division). Again, the space of the cosmos comes to mind, the space of the universe from infinite to infinitesimal, and of space-time, maps changing second by second.

The space where the sun almost touches Earth on a continuous uninterrupted cycle throughout the cosmos.

Contrast the space of the "real world" (as it is so often called), the one filled with emails and expectations. Like an astronaut returning from orbit and feeling once again the unrelenting insistence of Earth's gravitational pull, transitioning between these two distinct spaces requires a period of acclimation and adjustment moving from odds to evens and back again.

•••

For me as an artist, there are other spaces as well.

Visual space, where light adaptation occurs: the removal of all light and all objects so that the retina can reset and clarity of vision can be restored. This, too, can be metaphorical—as in the clarity of understanding and of ideas.

The space of reflection, where the retina and light waves moving at 300,000 kilometers per second magically connect.

The space of materiality, where exists pure appreciation and endless joy and fascination for materials, their unique physical properties and characteristics, their relationship to gravity and to light. For example, this space as it relates to transparent materials, how color passes through and appears in unexpected places. Or this space as it relates to reflective surfaces, how a mirrored infinity effect creates a reality that's only real to the observer at that moment, at that particular vantage point.

Transparent or translucent materials can share this space of movement where light passes through unique to the molecules of the mass.

The space where molecules bond in acrylic material and light waves pass through without hesitation or inhibition exploding in all directions.

The space of perspective, where angles, positions, distances change rapidly on courses more ancient than the human species.

- Shana Mabari
Artist

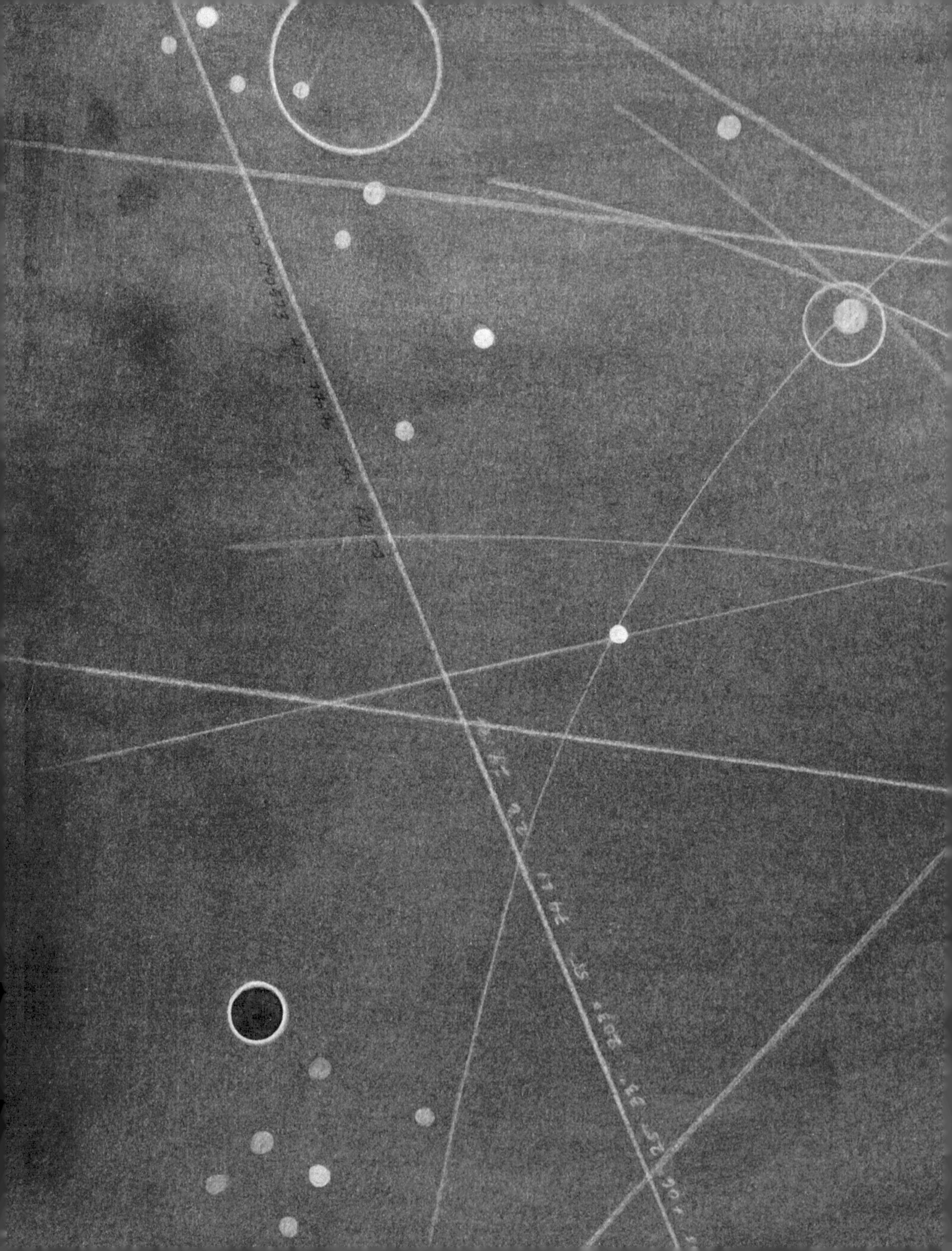

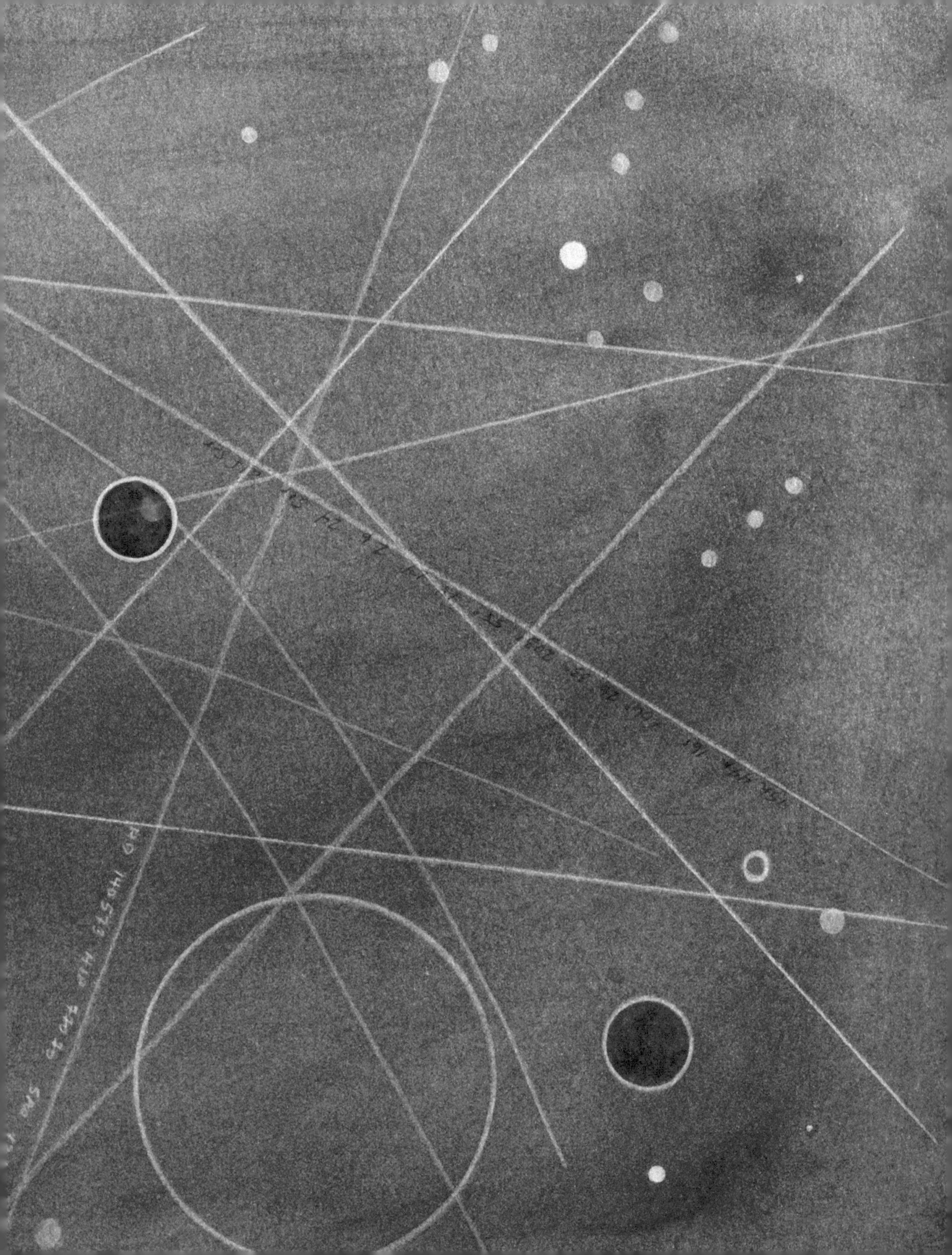

SPACE AND CONNECTION

In the beginning, the child is grown inside a liquid nebula. Without the nuisance of gravity. This space ticks with the song of creatures under a still sea. Fleshy walls seep the muted swell of voices. Light is dark or reddish as when eyes are closed against the sun. This space is no-thing, the last place where awareness is purely metaphysical. The womb is contained and vast, expanding and yet confined. It is the closest the child will ever be to the universe.

Everything aligns and, just like a star, the balance of forces compress enough to squeeze the child out. Suddenly there is nothing but space and unnamed sensations. She does not know the feeling of her skin against cloth, against latex gloves, against her mother's skin. At first, the infant does not differentiate herself from the mother. The child is a sun. At least that is how her marrow understands things to be. The mother is a moon. Space exists in the gaps of things, spreading into soft folds, into the crook of an arm, into the time it takes to answer a cry. As sensation gives way to the space the infant becomes a giant gaping mouth, opening, expanding, gulping life. Consuming it as if it is never-ending.

The infant grows into a child and the child swings in her spaces at times sure of everything and at times sure of nothing. The mother weaves in and out of the fabric of the child's experience. The child does not notice the folds and creases, the dark spaces where her origin story is threaded with that of the others. The spaces between weft and warp expand and contract. The child grows, as we know, not always elegantly. For the mother, the fabric is distorted. Those cozy folds and air pockets are not innocent. Those spaces are speckled with dust particles, remnants of interstellar events (reactive, radioactive, radiant). Embedded within the weft and warp is the mother's own mother and so on. Space becomes shared across dimensions, reaching back to where the threads are no longer visible. Past each mother in turn, until we arrive at the first mother.

The maw of separateness devours the thread that connected the child to the womb. To the knowledge of the universe. To her bodily wisdom. The steady orbit of childhood gives way to a careening, a flight path skewed, sometimes coming into land, sometimes not. The child grows into an adolescent. Space rushes into everything. It's between herself and all the others: parents, siblings, friends, lovers. There is even a murky space between what she thinks of her self and what she wants that self to be. This space stretches and snaps, goes slack then taut, twists and turns and like the universe is eternally expanding without her noticing. Until there is an interstellar event (reactive, radioactive, radiant), and she is exploded into a million shards of matter. Debris litters the spaces of life now, floating untethered before coming together again. The child grows into an adult, into a mother.

For the mother, the body is no longer a private space. It will never return to a time where it had its own story. The mother moves from a body whose spaces are known, understood, regarded, and defined to a space where the body is unidentified. It reveals new secrets as if all this time it was hiding things. Space is no longer a feature of the mother's life. She lives in spacelessness. Her body, her mind, her soul will be shared from now on no matter the trajectory of the child. Even a mother estranged or grieving lives in a shared space where no one can hear her cries even if they are cries of love.

Alongside this space is another. It is enormous. Larger than anything else that exists in this universe or the next. It expands at a velocity that overwhelms. This space is invisible, a private universe shared only with her infant. It is everywhere but nowhere. It is space beyond love. And yet the universe of the mother is confined. Here space becomes something to fight for. There is never enough of it. Obligations fill the once soft folds and creases of the once warm and welcoming fabric. Space is narrow and momentary. To capture it requires an exhausting awareness. A vigilance for the way the light falls on leaves, the ripples of a still sea, a perfectly made cup of tea. These moments suffice, become joy, mostly. These are her spaces now. She is never again alone with herself.

- Danni Landa
Writer

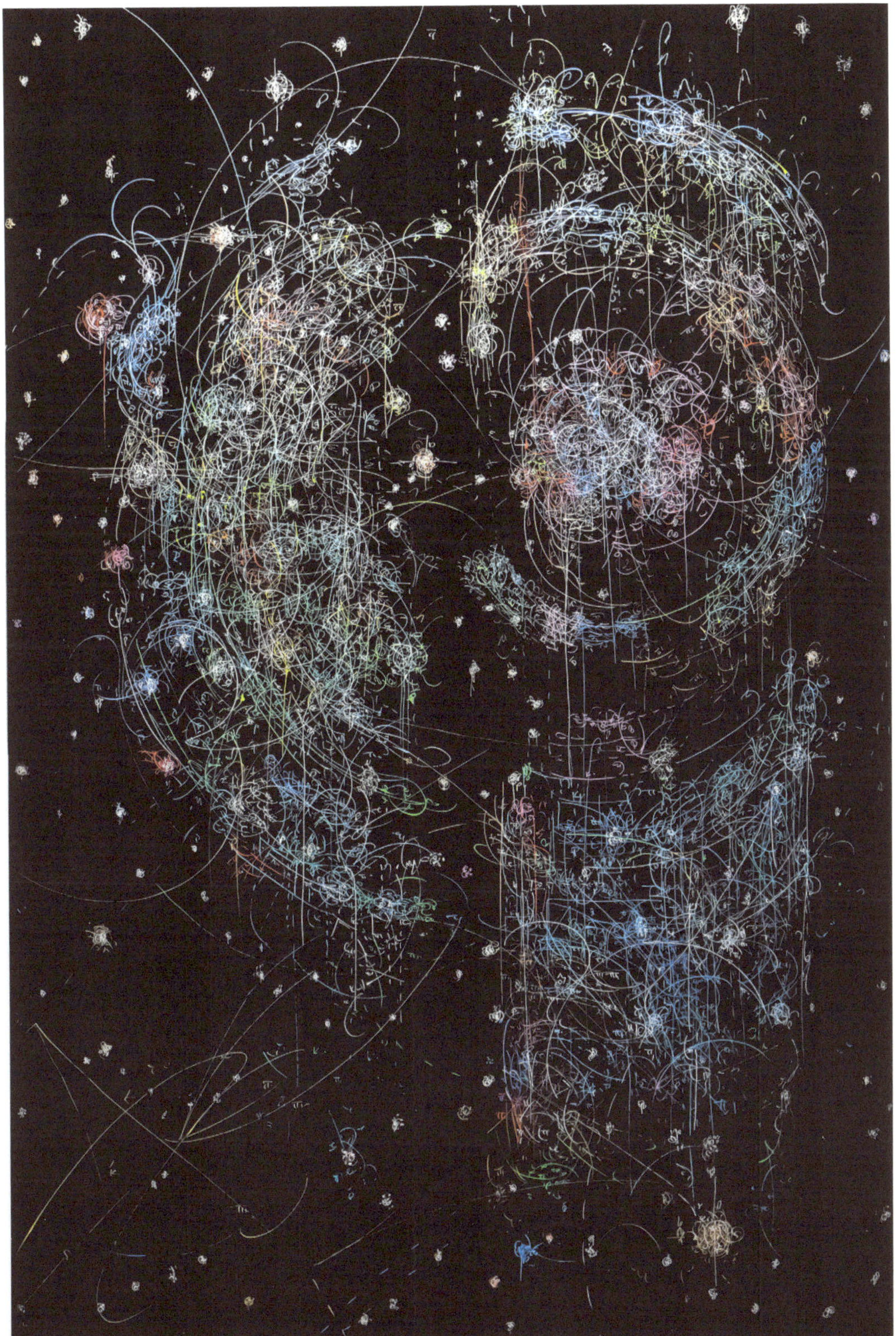

INDEX OF ARTWORKS

Cover
Shana Mabari
Unuk Serpens 74 Light-Years, 2019
Silver aluminum print,
20 × 30 in.

p. xv
Mary Anna Pomonis
She Is in the Radiant Light, 2019
Acrylic airbrush on canvas,
36 × 36 in.

p. xx
Shana Mabari
Mars – 6,779 km, 2019
Silver aluminum print,
20 × 30 in.

p. xxvi
Marcia Roberts
La Panza, 2002
Acrylic on canvas, 80 x 112 in.

p. 5
Robert Standish
Between Here and There, 2019
Acrylic on canvas,
56 in. (diam.)
Collection of Meg Joannides,
Los Angeles, CA

p. 11
Gary Lang
Gilded and Coal, 2009
Acrylic on canvas,
13 ft. (diam.)
Private collection

p. 21
Edwin Vasquez
Infinite Refraction, 2019
Digital image

p. 27
Shana Mabari
Astral Challenger, 2016
Stainless steel and acrylic,
20 × 16 × 16 ft.
Commissioned by the City of Lancaster, CA Photograph by Eric Staudenmaier

p. 31
Jeff Frost
Still from *GO.HOME*
2018 Video

p. 36
Alex Couwenberg
Billy Two Rivers, 2018
Acrylic and spray paint on canvas, 40 × 60 in.

p. 45
Laddie John Dill
Sand and Light, 2017
Silica sand and neon light,
variable dimensions

p. 51
Phillip K. Smith III
Lucid Stead, 2013
Wood, LED light, and mirror,
variable dimensions
Photograph by Lance Gerber

p. 55
Jay Mark Johnson
CITTA DEI MOTORINI #4,
2009
Durst Lambda print and
aluminum, 40 × 384 in.
Edition 1 of 3

p. 63
Casper Brindle
HOME, 2018
Acrylic and resin on panel,
60 × 60 in.

p. 71
Shana Mabari and
Dr. Shinsuke Shimojo
BLUE VOLUMETRIC, 2012
Acrylic, steam, blue light,
42 x 42 x 42 in.

p. 79
Shana Mabari
Meteor, 2019
Acrylic, 5 × 5 × 5 in.

p. 85
Ruth Pastine
Blue Darker Light, 2017
Oil on canvas on beveled
panel, 60 × 32 × 3 in.
Private collection, Newport
Beach, CA

p. 91
Shana Mabari
Unuk Serpens 74 Light-Years,
2019
Silver aluminum print,
20 × 30 in.

p. 97
Kysa Johnson
*Blowup 320 – the long good-
bye (history of gold) subatomic
decay patterns and binary
neutron star in Circinus X-1*,
2017
Ink on gloss on panel,
78 × 48 in.

SHANA MABARI

Shana Mabari is an American contemporary artist with a studio practice based in Los Angeles and Ibiza, Spain. Previously she has lived and worked in Paris, Northern India, Southeast Asia, and Tel Aviv. Central to Mabari's practice are her ongoing investigations of the intersections of art and science, as evidenced by her sculptures, installations, and immersive environments that explore the dynamics of visual perception and the ways in which we experience physical space. The artist's engagement with color, light, reflection, and geometric form can be situated on a continuum that connects to the Light and Space movement that originated in California in the 1960s, while her vision of an expanded practice in the twenty-first century centers on exploring how the space of art making and viewing might productively intersect with what are otherwise often highly technical and advanced scientific fields, such as astrophysics and psychophysics. In 2004, Mabari and Dr. Shinsuke Shimojo of the California Institute of Technology were awarded a patent for the design of *Dynamic Spatial Illusions*, a portable version of a visual-and-sensory experimental environment, and the artist has ongoing collaborations with experts in the vision sciences at Caltech and in the neurosciences at Zurich's Institute of Neuroinformatics. In 2018, Mabari was selected to be the first artist to fly aboard a mission of NASA's Stratospheric Observatory for Infrared Astronomy (SOFIA), which uses a 2.7-meter telescope mounted in a customized 747 flying to a maximum altitude of 45,000 feet to study such astronomical phenomena as black holes and star formation. It is through the artist's own intersections with advanced scientific fields that she explores the complexities of the contemporary individual's relationship to the ways in which these disciplines have dramatically expanded our collective field of vision and our understanding of physical reality, which in turn serves as a key animating force in her production.

ANDI CAMPOGNONE

Andi Campognone has over 30 years of arts experience in the southern California region. She is the Executive Director of AC Projects, a non-profit consulting organization focused on promoting arts and culture. Projects include developing museum exhibitions, public engagement, mentoring programs and book and film publications of historically relevant southern California artists. Campognone is also the Arts Manager/Senior Curator for the City of Lancaster's museums. She is responsible for the development and maintenance of partnerships and community engagement initiatives with artists, businesses, stakeholders, Los Angeles County Arts Commission, Los Angeles County Supervisors office, California Arts Council and higher level institutions. She develops curatorial direction for exhibition, educational and engagement programming and additionally leads the public art initiatives for the City of Lancaster. She has previously served the City of Pomona as Cultural Arts Commissioner where she co-wrote and implemented the City's Master Cultural Arts Plan and the adopted Arts in Public Places Policy. She is the cofounder of Kipaipai, a professional development workshop for artists in Hawaii, California and New York. Campognone is on the Board of the Lancaster Museum and Public Art Foundation and on the Board of the Holualoa Foundation for Arts and Culture. She volunteers as a regular speaker and mentor to art students at both the undergraduate and graduate level and is on the advisory board of the Los Angeles Arts Association. She is a member of ArTTable.

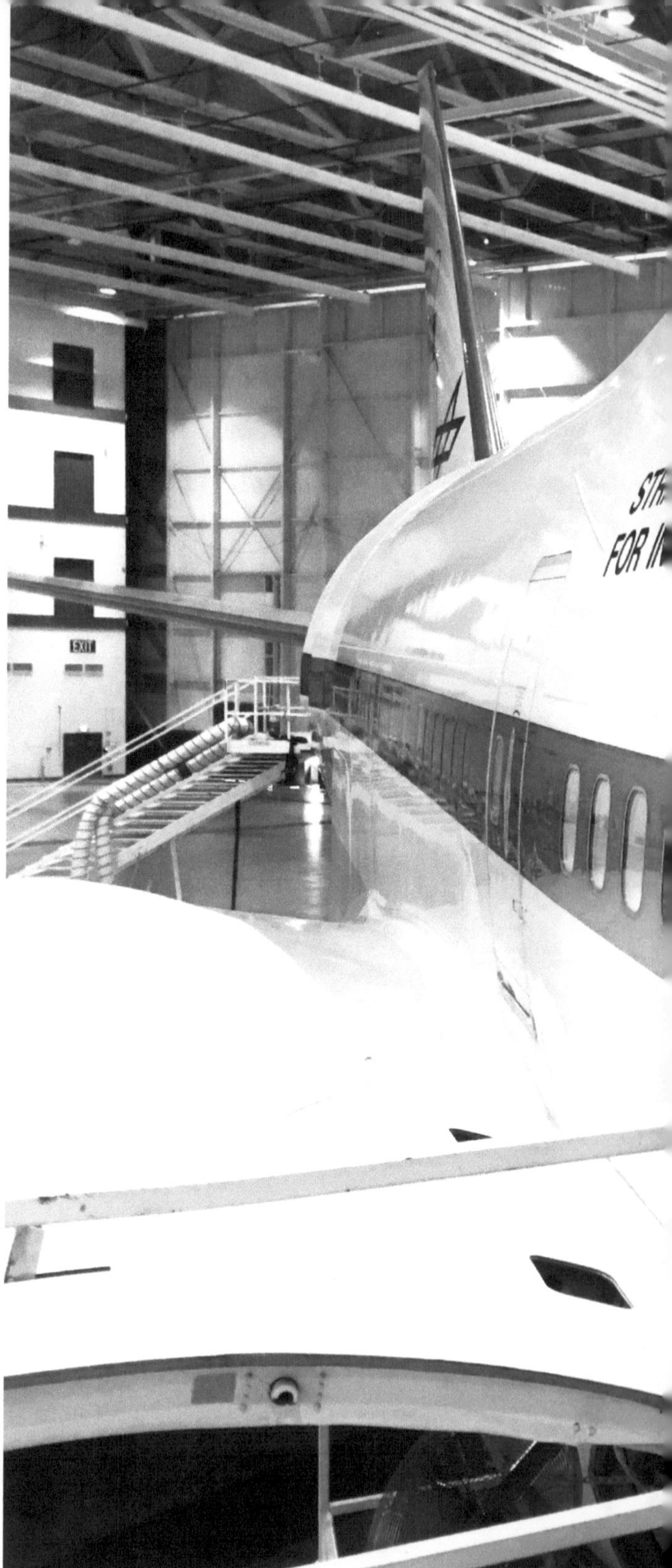
EXIT

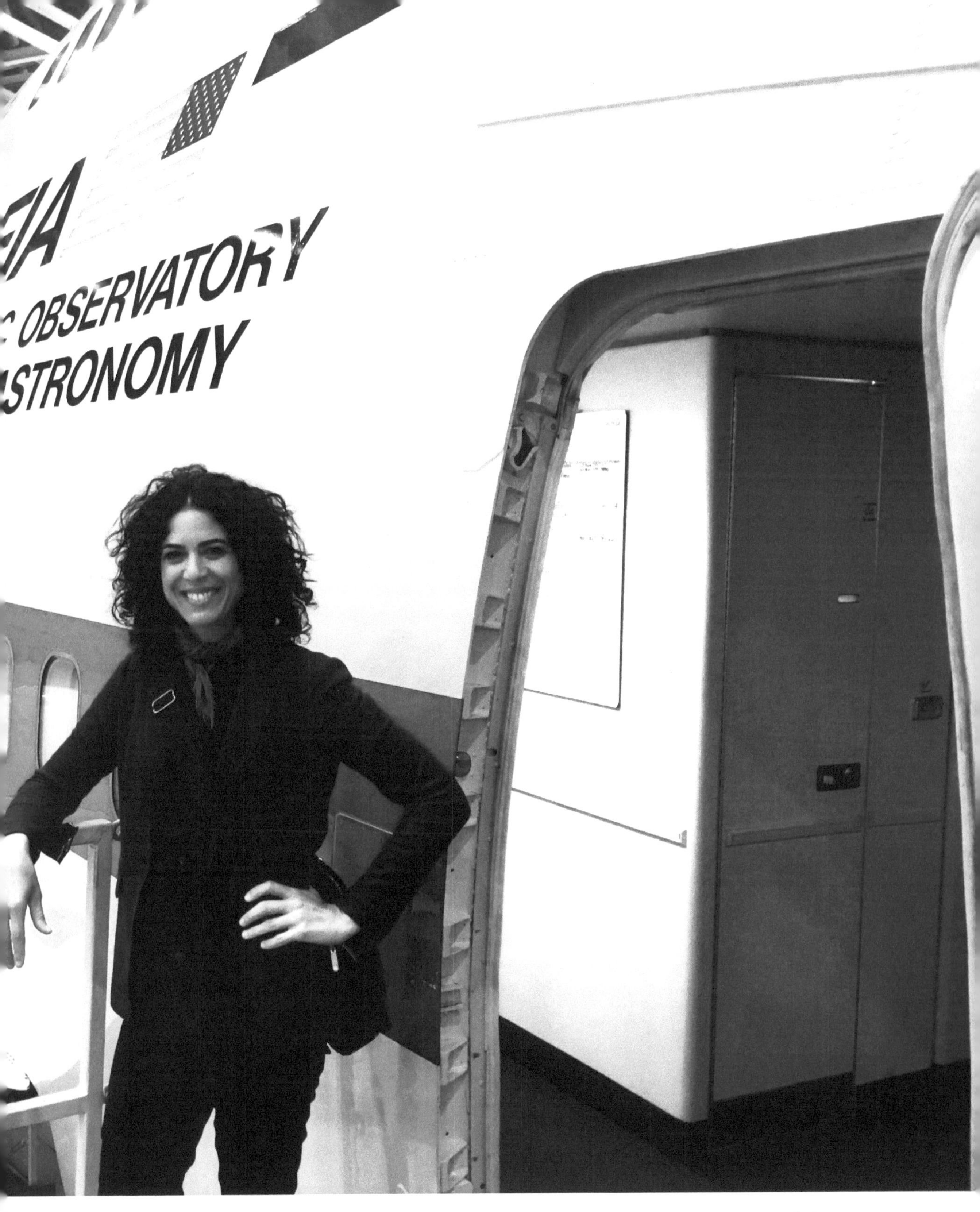

Shana Mabari / SOFIA flight 2018.
Image courtesy of NASA Armstrong